Conference Proceedings of the Society for Experimental Mechanics Series

Series Editors

The Conference Proceedings of the Society for Experimental Mechanics Series presents early findings and case studies from a wide range of fundamental and applied work across the broad range of fields that comprise Experimental Mechanics. Series volumes follow the principle tracks or focus topics featured in each of the Society's two annual conferences: IMAC, A Conference and Exposition on Structural Dynamics, and the Society's Annual Conference & Exposition and will address critical areas of interest to researchers and design engineers working in all areas of Structural Dynamics, Solid Mechanics and Materials Research.

Vijaya Chalivendra • Frank Gardea
Editors

Mechanics of Composite, Hybrid & Multi-functional Materials, Volume 5

Proceedings of the 2022 Annual Conference on Experimental and Applied Mechanics

Springer

Editors
Vijaya Chalivendra
SENG-116
University of Massachusetts
North Dartmouth, MA, USA

Frank Gardea
U.S. Army Research Laboratory
Katy, TX, USA

ISSN 2191-5644 ISSN 2191-5652 (electronic)
Conference Proceedings of the Society for Experimental Mechanics Series
ISBN 978-3-031-17444-5 ISBN 978-3-031-17445-2 (eBook)
https://doi.org/10.1007/978-3-031-17445-2

This Springer imprint is published by the registered company Springer Nature Switzerland AG
The registered company address is: Gewerbestrasse 11, 6330 Cham, Switzerland

Preface

Mechanics of Composite, Hybrid and Multifunctional Materials represents one of six volumes of technical papers to be presented at the 2022 SEM Annual Conference & Exposition on Experimental and Applied Mechanics organized by the Society for Experimental Mechanics scheduled to be held, June 13–16, 2022. The complete Proceedings also includes volumes on: *Dynamic Behavior of Materials*; *Challenges in Mechanics of Time-Dependent Materials & Mechanics of Biological Systems and Materials*; *Fracture, Fatigue, Failure and Damage Evolution*; *Advancements in Optical Methods, Digital Image Correlation & Micro- and Nanomechanics*; and *Thermomechanics & Infrared Imaging, Inverse Problem Methodologies and Mechanics of Additive & Advanced Manufactured Materials.*

The commercial market for composite continues to expand with a wide range of applications from sporting equipment to aerospace vehicles. This growth has been fueled by new material developments, greater understanding of material behaviors, novel design solutions, and improved manufacturing techniques. The broad range of applications and the associated technical challenges require an increasingly multidisciplinary and collaborative approach between the mechanical, chemical, and physical sciences to sustain and enhance the positive impact of composites on the commercial and military sectors.

New materials are being developed from recycled source materials, leading to composites with unique properties and more sustainable sources. Existing materials are also being used in new and critical applications, which require a deeper understanding of material behaviors and failure mechanisms on multiple lengths and time scales. In addition, the unique properties of composites present many challenges in manufacturing and in joining with other materials. New testing methods must be developed to characterize the novel composite properties, to evaluate application and product life cycle performance, as well as to evaluate impacts and merits of new manufacturing methods.

This volume presents early research findings from experimental and computational investigations related to the processing, characterization, and testing of composite, hybrid, and multifunctional materials.

North Dartmouth, MA, USA — Vijaya Chalivendra
Katy, TX, USA — Frank Gardea

Contents

1 **Carbon Fiber Reinforced Polymers with Carbon Nanotubes: Investigation of Interlaminar Strength** 1
Andres Argenal, David Matthews, Connor Murrell, Andrew H. Cannon, Mark Pankow, and Garrett J. Pataky

2 **Recycled Natural Rubber-Based Composites Reinforced with Nano Boron Nitride in Thermal Conductive and Electrical-Insulating Fields** 7
Gamze Çakır Kabakcı, Matthieu Sonar, Özgür Aslan, and Emin Bayraktar

3 **Impact Behaviour of Recycled Rubber-Based Composites Reinforced with Glass Bubbles and Alumina Fibers (γ-Al_2O_3)** 17
Gamze Çakır Kabakcı, Özgür Aslan, and Emin Bayraktar

4 **Toughening Mechanism of Silicon Whiskers and Alumina Fibres (γ-Al_2O_3) Reinforced Ni-Al-Cu Matrix Composites Through "Sintering + Forging"** 29
Ibrahim Miskioglu, Georges Zambelis, Fabio Gatamorta, Ozgur Aslan, and Emin Bayraktar

5 **Multifunctional Behaviour of Titanium-Based Composites Reinforced with Recycled Iron and Magnetic Iron Oxide (Fe_3O_4): Magnetic Shape Memory Composites** 37
Stanislav Slavov, Ivelina Ivanova, Virginia Alexandrova, Dimitar Dontchev, and Emin Bayraktar

6 **Recycled Ti-Al-Cu Matrix Composites Reinforced with Silicon Whiskers and γ-Alumina (Al_2O_3) Fibres Through Sintering + Forging** 49
Fabio Gatamorta, Ibrahim Miskioglu, Dhurata Katundi, and Emin Bayraktar

7 **Development of Ni-Al/Nb_2Al/ZrO_2-Based Composites for Aircraft Engine Applications Produced by a Combined Method: Sintering + Forging** 55
Utku Kaftancıoglu, Georges Zambelis, Fabio Gatamorta, Ibrahim Miskioglu, and Emin Bayraktar

8 **Out-of-Plane Restraint Under Tension In-Plane Loading** 61
Dawn C. Jegley, W. Allen Waters Jr., and Alana Zahn Cardona

9 **Preconditioning Layers Affect Osteoblastic Cell Adhesion to Orthopedic Implant Surfaces** 69
James D. Boyd and Martha E. Grady

10 **Development of a Polyamide 6-Based Composite Material for UAV Propellers** 73
Aybeniz Seyhan, Alaeddin Burak Irez, and Yusuf Polat

11 In Situ Measurement of AC Conductivity to Quantify Unidirectional Alignment of Graphene Nanoplatelets (GNPs) in Epoxy 79
Sai Tharun Kotikalapudi and Raman P. Singh

12 How Graphene Oxide Content in Poly (Ethylene Glycol) Affects Phase Separation in Epoxy Matrix for High-Performance Structural Supercapacitor Applications 87
Alan J. Austin, Sai Tharun Kotikalapudi, Padmanapan Rao, and Raman P. Singh

Chapter 1
Carbon Fiber Reinforced Polymers with Carbon Nanotubes: Investigation of Interlaminar Strength

Andres Argenal, David Matthews, Connor Murrell, Andrew H. Cannon, Mark Pankow, and Garrett J. Pataky

Abstract The interlaminar fracture of carbon fiber reinforced composites is a leading cause of failure while in service. In an effort to overcome this weakness, carbon nanotubes were embedded into HexPly® IM7/8552 Prepreg in an attempt to increase the interlaminar strength. The present work experimentally analyzed the mechanical properties of the composite materials with and without carbon nanotubes. A systematic curing process was designed to fabricate composite laminates in either an 8-ply quasi-isotropic or 12-ply unidirectional configurations. Property variations at 0°, 30°, 45°, 60°, and 90° were investigated using the unidirectional laminates. Samples were subjected to supersonic impact and 4-point bending tests. Scanning electron microscopy was used to analyze the postmortem specimens. The 4-point bending tests were analyzed using digital image correlation to measure displacement, strain fields, and calculate flexural strength.

Keywords Carbon fiber · Carbon nanotubes · Interlaminar strength · Impact · Delamination

1.1 Introduction

Carbon fiber reinforced polymer composites are a supplement for traditional metallic alloys, such as aluminum in aerospace applications. These materials allow a reduction in weight while maintaining similar structural strength. A tradeoff exists while using carbon fiber materials since composites have low interlaminar strength and are susceptible to delamination. The harsh environment aircraft components endure increases the failure modes for composites [1]. A common cause of failure for these components develops from impact damage. Impact damage can lead to several failure modes in carbon fiber laminates, including delamination, debonding, fiber breakage, matrix cracking, and a mixture of each failure mode [2]. The purpose of this study is to investigate if prepreg carbon fiber reinforced polymers embedded with carbon nanotubes developed by N12 Technologies, Inc. increases the mechanical properties. Figure 1.1 shows the carbon nanotubes were orthogonal to the fibers to increase the shear strength and fracture toughness in regions where composites experience delamination [3].

The carbon nanotubes were embedded into Hexply® IM7/8552 prepreg. IM7/8552 is an aerospace-grade carbon fiber reinforced polymer composite with 12,000 filament count tows. The fibers are surface treated for prepregging applications producing a tensile strength of 5655 MPa and tensile modulus of 276 GPa. N12 Technologies claimed that the addition of carbon nanotubes increased the interlaminar strength of these composites by +5% [3].

A. Argenal · D. Matthews · C. Murrell · G. J. Pataky (✉)
Department of Mechanical Engineering, Clemson University, Clemson, SC, USA
e-mail: aargena@clemson.edu; dsmatth@clemson.edu; cmurrel@clemson.edu; gpataky@clemson.edu

A. H. Cannon
1900 Engineering LLC, Clemson, SC, USA

Department of Chemical and Biomolecular Engineering, Clemson University, Clemson, SC, USA
e-mail: ahcanno@clemson.edu

M. Pankow
Department of Mechanical and Aerospace Engineering, North Carolina State University, Raleigh, NC, USA
e-mail: mrpankow@ncsu.edu

V. Chalivendra, F. Gardea (eds.), *Mechanics of Composite, Hybrid & Multi-functional Materials, Volume 5*, Conference Proceedings of the Society for Experimental Mechanics Series, https://doi.org/10.1007/978-3-031-17445-2_1

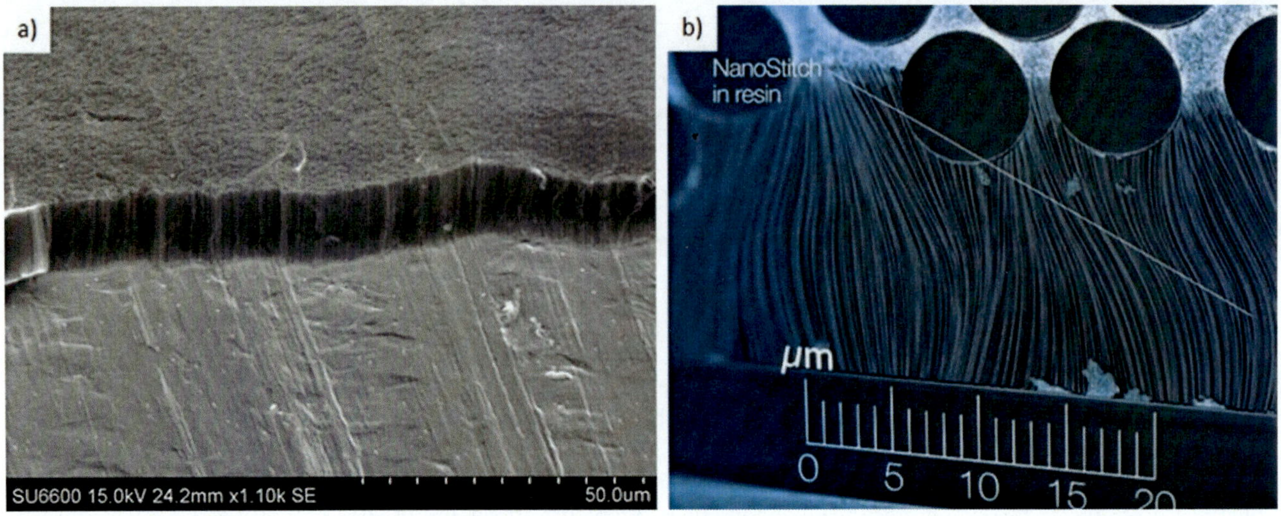

Fig. 1.1 (**a**) SEM image of vertically aligned carbon nanotubes on a metal substrate. (**b**) Cross section of carbon nanotubes interfacing with carbon fiber prepreg [4]

1.2 Background

Specimens were produced in cross-plied quasi-isotropic and unidirectional configurations, shown in Fig. 1.2. The quasi-isotropic configuration consisted of 8 plies with a balance orientation of 0°/90°/+45°/−45°/−45°/+45°/90°/0° to achieve symmetry. The unidirectional configuration consisted of 12 plies in the same orientation and was studied in different angles, including 0°, 30°, 45°, 60°, and 90° variations.

The laminates were produced using a vacuum bagging technique where the prepreg composites were vacuum sealed and preheated at 177 °C for 10 min. The plate was removed from the bag and subjected to a press cycle. A heated press was used at 168 °C with a pressure of 192 MPa for 5 min to eliminate voids within the layup. The curing process was finalized by placing the laminate inside an oven at 177 °C for 3 h. The temperatures used were the same as those that Hexcel recommends for an autoclaving process [5].

The manufactured laminates, both with and without carbon nanotubes, were subject to high-velocity impact tests, performed at the BLAST lab at North Carolina State University. Impact tests were conducted using a ping-pong supersonic cannon, shown in Fig. 1.3, with an average velocity of 390 m/s.

Three quasi-isotropic specimens and seven unidirectional specimens were tested with variations in eight and twelve plies, listed in Table 1.1. All plates tested were square with a minimum size of 101 mm × 101 mm. Variations in size were due to excess material post-layup process. The laminates were not machined to the same size to avoid machining-induced delamination or damage. For the impact experiments, the specimens were clamped on all four sides onto the test fixture to maintain consistency for the exposed region.

The second series of experiments subjected the laminates to 4-point bending tests using the D6272 ASTM procedure to determine the strength difference between the laminates with and without carbon nanotubes. Following the ASTM procedure, a support length of 25.4 mm was established. The specimens were produced from a 305 mm by 305 mm laminate by waterjet cutting out specimens with a width of 12.3 mm and a thickness of 2.04 mm. 4-point bending experiments were conducted in a servo-hydraulic load frame with a set displacement rate. The experimental setup for the 4-point bending tests is shown in Fig 1.4. The 4-point bending tests consisted of two 0° unidirectional without carbon nanotubes specimens and one 0° unidirectional specimen with carbon nanotubes embedded. Deformation was analyzed using digital image correlation (DIC) to measure displacement and identify the onset of failure during the experiments. The edge of the specimens was speckled via airbrush before the experiments for DIC. The flexural strength (S) at failure was obtained using Eq. 1.1 [6], where P is the load, L is the length of support span, b is the specimen's thickness, and d is the specimen's width.

$$S = PL / bd^2 \tag{1.1}$$

The deformation analysis also included finding the energy produced on the specimens. After determining the flexural strength, the energy was found using the definite integral of a high degree polynomial obtained from force versus displacement data and using Eq. 1.2, where ε_f is the strain upon failure and σ is the stress (Fig. 1.4).

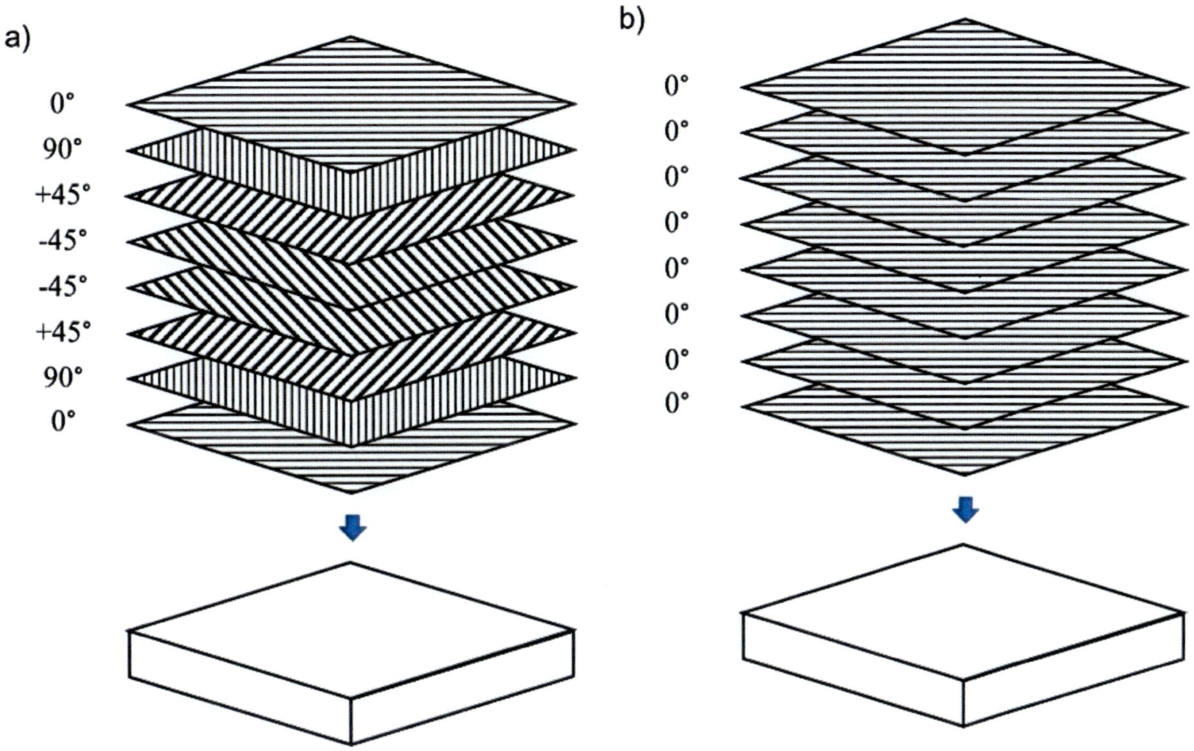

Fig. 1.2 Schematics of the (**a**) quasi-isotropic layup and (**b**) unidirectional layup

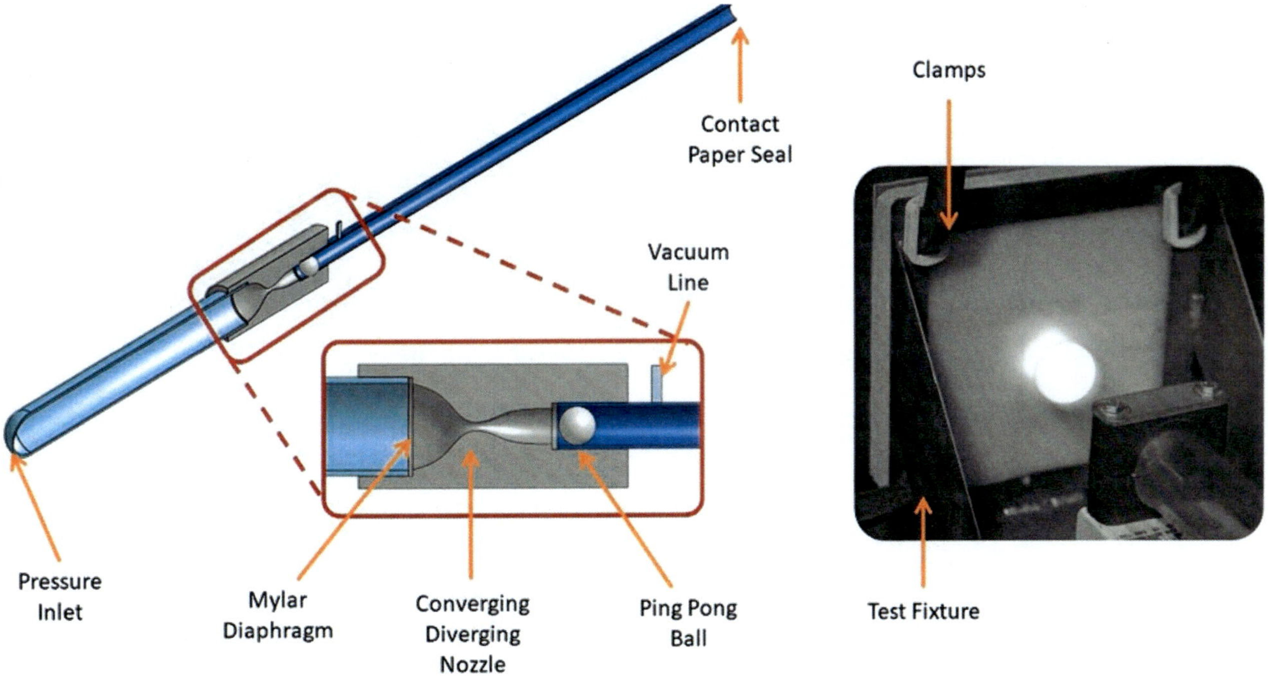

Fig. 1.3 Supersonic ping-pong cannon

$$\text{Toughness} = \frac{\text{energy}}{\text{volume}} = \int_0^{\varepsilon_f} \sigma d\varepsilon \tag{1.2}$$

Table 1.1 Fiber orientation, composite type, ply numbers, and geometrical constraints of specimens used in impact experiments

Fiber orientation	Composite type	# of plies	Length (mm)	Height (mm)	Thickness (mm)
[0°]	IM7 + carbon nanotubes	12	101.6	101.6	23.8
[0°]	IM7 + carbon nanotubes	8	101.6	101.6	16.7
[0°]	IM7 + carbon nanotubes	12	101.6	101.6	24.1
[0°]	IM7 + carbon nanotubes	8	101.6	101.6	16.7
[0°]	IM7 + carbon nanotubes	8	101.6	101.6	17.2
[0°]	IM7 + carbon nanotubes	8	101.6	101.6	16.0
Quasi-isotropic	IM7 + carbon nanotubes	–	101.6	101.6	17.5
Quasi-isotropic	IM7 + carbon nanotubes	–	101.6	101.6	17.5
[0°]	IM7	12	152.4	152.4	22.6
Quasi-isotropic	IM7	–	152.4	152.4	18.0

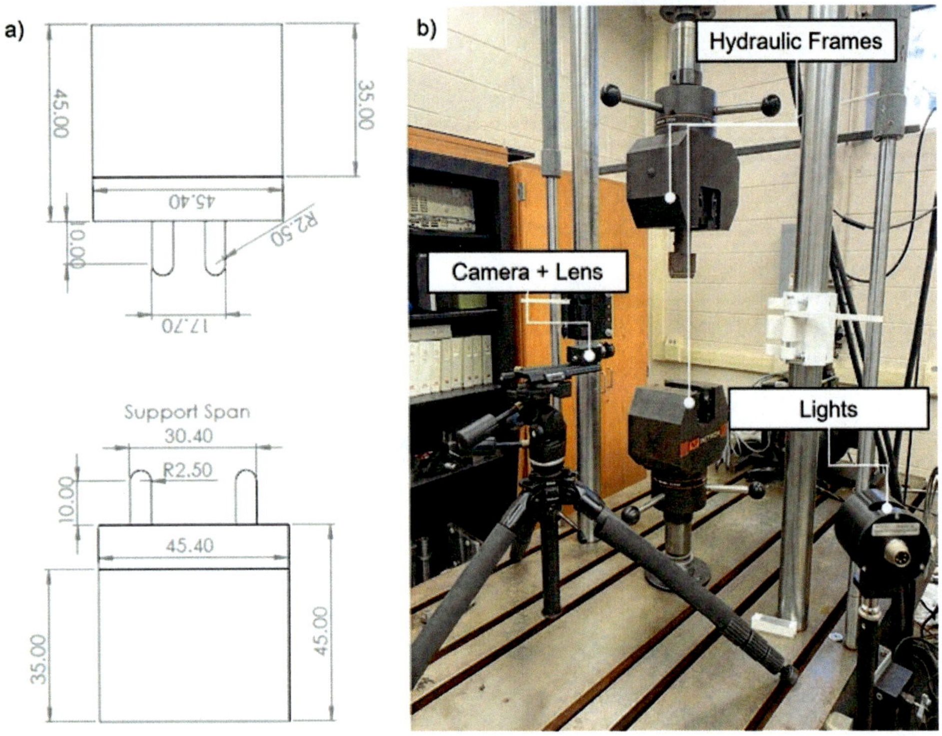

Fig. 1.4 (**a**) Schematic of the 4-point bending grips. (**b**) Load frame and experimental setup for DIC

1.3 Analysis

1.3.1 Impact Experiments

The impact experiments were recorded with a high-speed camera to quantify the ping pong ball velocity and qualitatively analyze the laminate behaviors, with an example impact sequence shown in Fig. 1.5. Postmortem specimens were analyzed from their structural and delamination performance. Fractures and cracks determined the structural performance of the specimens. Delamination was investigated from broken fibers produced by impact, fractures, or cracks, and was quantified as single or multiple depending on the severity of fiber misalignment, with the full results given in Table 1.2. The specimens that experienced significant structural and delamination damage were the 0° unidirectional specimens with carbon nanotubes. In severe cases, they had complete fractures and multiple delaminations. The control IM7 specimens, both quasi-isotropic and 0° unidirectional specimens, were able to sustain the impact test resulting in no fractures and cracks, with single delamination. The quasi-isotropic specimen with carbon nanotubes matched the performance of the 12-ply, 0° unidirectional specimen without carbon nanotubes producing the best-performing specimens. Both specimens experienced no visible cracking, and minimal delamination was present at the impact region (Fig. 1.5 and Table 1.2) indicated earlier in paragraph now.

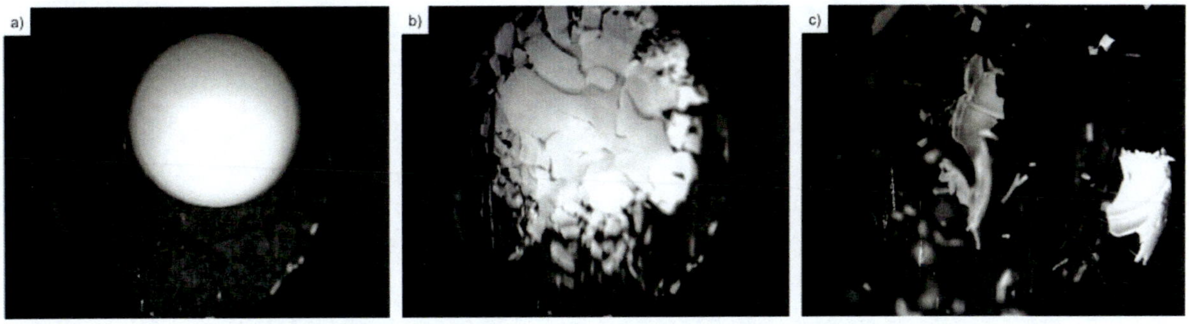

Fig. 1.5 Quasi-isotropic specimen without nanotubes (**a**) before impact, (**b**) during impact, and (**c**) immediately post-impact

Table 1.2 Velocity data, structural analysis, delamination analysis, and velocity data of the impact experiments

Fiber orientation	Composite type	# of plies	Velocity (m/s)	Fracture	Delamination
[0°]	IM7 + carbon nanotubes	12	423	Complete fracture	Multiple
[0°]	IM7 + carbon nanotubes	8	423	Complete fracture	Multiple
[0°]	IM7 + carbon nanotubes	12	397	Single crack at impact site	None identified
[0°]	IM7 + carbon nanotubes	8	397	Minimal cracking at impact site	Single
[0°]	IM7 + carbon nanotubes	8	397	Two cracks	None identified
[0°]	IM7 + carbon nanotubes	8	346	Multiple cracks	Multiple
Quasi-isotropic	IM7 + carbon nanotubes	–	397	No visible cracks	Single
Quasi-isotropic	IM7 + carbon nanotubes	–	423	Minimal cracking at impact site	Single
[0°]	IM7	12	282	No visible cracks	Single
Quasi-isotropic	IM7	–	318	No visible cracks	Single

1.3.2 4-Point Bending Experiments

The center section between the two upper contact points was analyzed with DIC using a sample rate of 5 Hz. The images and the DIC analysis were used to identify the onset of failure. An example is shown in Fig. 1.6 for the delamination and cracking of the 0° unidirectional 12-ply specimen without carbon nanotubes. The load at the time each image was captured was recorded, thus allowing for it to be used in the flexural strength analysis, Eq. 1.1. Specimens without carbon nanotubes had an average flexural strength of 783.25 MPa, while the specimens with carbon nanotubes had a lower flexural strength of 565.4 MPa. The toughness was found using Eq. 1.2. Specimens without carbon nanotubes had a toughness average of 96.8 J m^{-3}, while the specimen with carbon nanotubes had a higher toughness of 133.9 J m^{-3}. The results from the 4-point bending experiments are shown in Table 1.3. Although the strength was found to be lower, defects due to manufacturing may have played a role in the earlier failure. The increased toughness, despite the lower flexural strength, indicates that the carbon nanotubes show promise for increasing the interlaminar strength. In all three specimens tested, delamination between the inner plys was identified as the onset of failure. Thus the increased toughness was concluded to be a delay in the plies separating. However, with such a low number of specimens tested, conclusive evidence on the impact of the carbon nanotubes cannot be made.

1.4 Conclusion

Multiple samples of both control and experimental specimens were manufactured through an alternative curing process, increasing the experiments' repeatability. From the high-velocity impact experiments, the quasi-isotropic specimens with carbon nanotubes specimens matched the structural and delamination performance of both the 0° unidirectional 12-ply and quasi-isotropic without carbon nanotubes showing minimal improvement with the embedded carbon nanotubes. The 4-point bending experiments concluded a decrease in flexural strength with embedded carbon nanotubes from the control specimens by 27.8%. However, the toughness increased by +38.3%, indicating a delay in delamination.

Acknowledgments Support for this project was provided by Clemson University's Creative Inquiry Program. The authors thank Hexcel Corporation and N12 Technologies, Inc for providing the carbon fiber prepreg material used in this study.

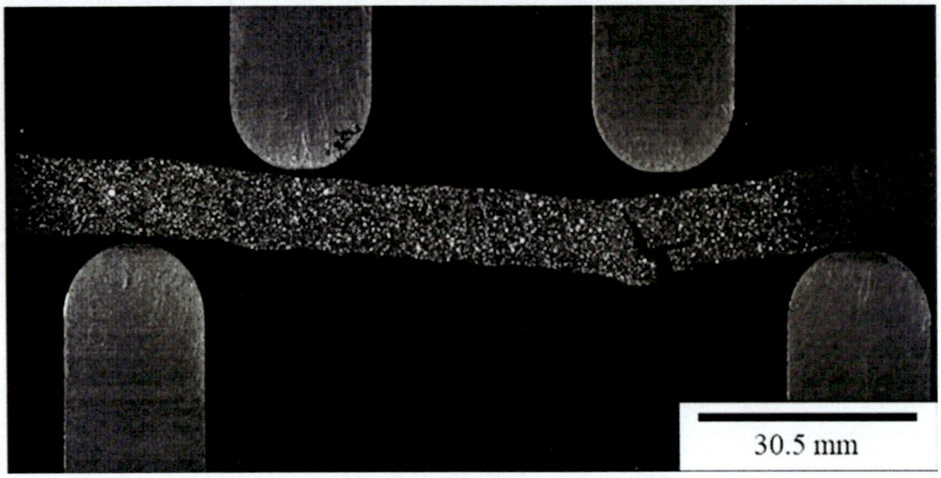

Fig. 1.6 Image at the onset of failure of the 0° unidirectional 12-ply 4-point bending specimen without carbon nanotubes.

Table 1.3 Flexural strength at failure, toughness, and work of 4-point bending experiments

Fiber orientation	Composite type	# of plies	Flexural stress (MPa)	Toughness (J m^{-3})
[0°]	IM7	12	809.4	98.9
[0°]	IM7	12	757.1	94.6
[0°]	IM7 + carbon nanotubes	12	565.4	133.9

References

1. Landry, B., Laplante, G., Leblanc, L.R.: Environmental effects on mode II fatigue delamination growth in an aerospace grade carbon/epoxy composite. Compos. A. Appl. Sci. Manuf. **43**(3), 475–485 (2012). https://doi.org/10.1016/J.COMPOSITESA.2011.11.015
2. Neidigk, S.: Detection and characterization of impact damage in carbon fiber aircraft fuselage structure. Master's thesis, The University of New Mexico (2013)
3. Gouldstone, C., Degtiarov, D., Williams, R.D.: Reinforcing plfy drop interfaces using vertically-aligned carbon nanotube forests. In SAMPE Conference Proceedings. Seattle, WA, June 2–5, 2014. Society for the Advancement of Material and Process Engineering (2014).
4. Conway, H., Chebot, D., Gouldstone, C., Williams, R.: Fatigue response of carbon fiber epoxy laminates with vertically-aligned carbon nanotube interfacial reinforcement. In SAMPE Conference Proceedings. Baltimore, MD, May 18–21, 2015. Society for the Advancement of Material and Process Engineering (2015)
5. Hexcel Corporation.: HexPly ® Prepreg Technology Datasheet (2013)
6. ASTM Standard D6272-17.: Standard Test Method for Flexural Properties of Unreinforced and Reinforced Plastics and Electrical Insulating Materials by Four-Point Bending. ASTM International, West Conshohocken, PA (2019). https://doi.org/10.1520/D6272-17

Chapter 2
Recycled Natural Rubber-Based Composites Reinforced with Nano Boron Nitride in Thermal Conductive and Electrical-Insulating Fields

Gamze Çakır Kabakcı, Matthieu Sonar, Özgür Aslan, and Emin Bayraktar

Abstract In this research, recycled natural rubber (NR) based composites reinforced with the doped nano boron nitride (NBN) with the high resolution in thermal conductive and electrical-insulating field were designed, and their properties were studied. Good distribution of doped NBN in the matrix has shown substantial increments of thermal conductivity and high electrical insulation depending on the quantity in the matrix. Additionally, the thermal analysis indicates that NR/NBN composites have excellent heat-transfer capacity during heating and cooling processes, which suggests great potential application in thermal conductive and electrical insulating fields. The procedure can find multiscale particle-matching ways to achieve the maximum effective thermal conductivity under a given filler load. It should be emphasized that the optimized effective thermal conductivity obviously can be improved with the increase in the volume fraction of the reinforcement.

Keywords Recycled natural rubber · Nano boron nitride · Thermal conductivity · Electrical-insulting field

2.1 Introduction

Manufacturing of the industrial parts from low-cost devulcanized recycled rubber-based composites are very economical and useful engineering applications to offer multifunctionality to the recycled constituent composites. In this study, nano boron nitrite (hBN) is favored as one of major reinforcements because it leads to perfect weight reduction, performance, processing and dimensional stability characteristics when used with other reinforcing fillers. Certain structural parts, in aeronautical applications, require high thermal conductivity capacity for these composites [1–8].

In the literature, high thermally conductive polymeric composites are studied by random mixing of thermally conductive and electrically insulated fillers with various polymers [2, 5–8]. The aim of this current study is to design recycled natural rubber-based composites reinforced with NBN to develop high thermally conductive and electrically resistive composites with suitable mechanical properties.

Boron nitride (BN) is a high thermally conductive and high electrically resistive material. It exists in various crystalline forms and the hexagonal form corresponding to graphite is the most stable and softest among BN polymorphs. hBN has a layered structure like graphite and shows anisotropic thermal properties [2, 5–8]. Thermal conductivity of perfect single

G. Çakır Kabakcı
Department of Mechanical Engineering, Atilim University, Ankara, Turkey

School of Mechanical and Manufacturing Engineering, ISAE-Supmeca-Paris, Paris, France
e-mail: gamze.cakir@atilim.edu.tr

M. Sonar · E. Bayraktar (✉)
School of Mechanical and Manufacturing Engineering, ISAE-Supmeca-Paris, Paris, France
e-mail: matthieu.sonar@isae-supmeca.fr; emin.bayraktar@isae-supmeca.fr

Ö. Aslan
Department of Mechanical Engineering, Atilim University, Ankara, Turkey
e-mail: ozgur.aslan@atilim.edu.tr

V. Chalivendra, F. Gardea (eds.), *Mechanics of Composite, Hybrid & Multi-functional Materials, Volume 5*, Conference Proceedings of the Society for Experimental Mechanics Series, https://doi.org/10.1007/978-3-031-17445-2_2

crystal hBN platelet is about 600 (W/mK) in the plane direction and 30 (W/mK) in out-of plane direction [2, 5–8]. hBN powders are in the form of single crystal platelets or spherical agglomerates.

Recycled natural rubber-based composites reinforced with the doped NBN, with the high resolution in thermal conductive and electrical insulating fields, are designed and mechanical properties are studied for seven different compositions in various weight percentages. In the frame of a common research with the French Aeronautical Society, a special work has been carried out for this study by using hBN in the recycled rubber-based composites. High thermal conductivity of the hBN should be useful for those composites that are aimed in certain parts of aircraft construction.

2.2 Design of Rubber-hBN Composite Type: Materials and Experimental Conditions

Fresh scrap rubber parts coming from sportswear factories are taken directly from production scraps and the first stage at the factory was undergone as a form of granule particles that after that are special. For manufacturing of recycled rubber-based composite, a good quality of bonding between matrix and reinforcements is required; however, recycled rubbers do not carry any free chains to form new bonding with epoxy resin because of the previous vulcanization process.

To improve the interface quality between epoxy and rubber, fresh scrap rubber should be devulcanized. During the devulcanization process, sulphur links are attempted to be broken and new other links are generated; then the flowing capacity and interaction of recycled rubbers with other substances are increased. After determination of the reinforcements with a matrix in wt%, a special process is applied to complete the manufacturing of the composites successfully. This process involves silanization of the recycled rubber and devulcanization before blending it with epoxy resin and reinforcement [1, 3, 4].

Chemical treatment follows devulcanized of recycled rubber and is mixed with pure epoxy resin powder to obtain a solid chemical bonding diffusion that is then strengthened with various reinforcements such as hBN powder (VWR—80% $d < 1$ µm), as presented the present study. After blending of the mixture, they are milled for 4 h by adding "Zn-Stearate" as a lubricant for homogenous distribution. For manufacturing of the composites, hot compaction is carried out at 220 °C for 20 min under the pressure of 6 tons; after that the specimens are produced in several diameter sizes (e.g., 30 mm, 50 mm, and so on). This process is given schematically in Fig. 2.1.

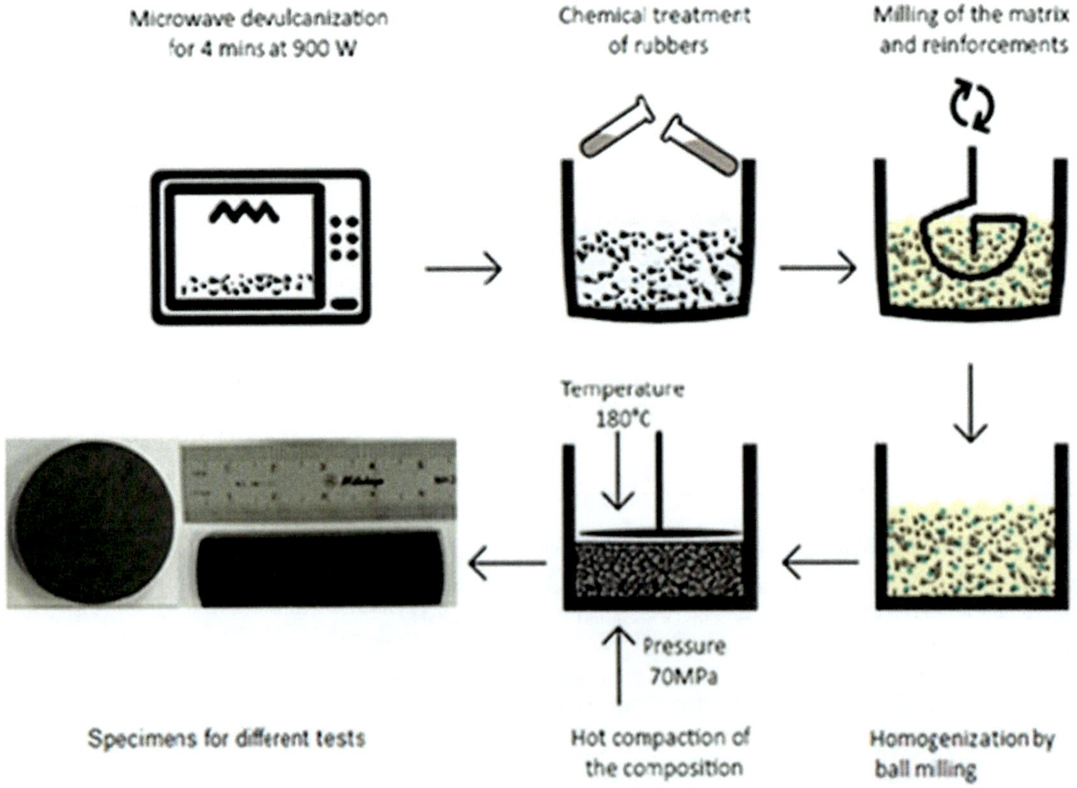

Fig. 2.1 Manufacturing steps of recycled rubber-based composites [1, 3, 4]

After manufacturing of the composites, different specimen geometries are prepared by using waterjet cutting for mechanical tests (e.g., compression, 3P-bending, impact test, and so on). Figure 2.2 shows the hBN powder as the reinforcement for rubber powder and some samples obtained from the mixed powder.

In the same way, a three-point bending (3 PB) test is conducted as shown in Fig. 2.3. This test device was adapted on the Zwick test apparatus and microstructure of the sample (BN-I) taken with a scanning electron microscope (SEM). Deflection of the specimen was measured by the crosshead position. Flexural strength, modulus of elasticity in bending and strain are obtained from the test results.

Microstructure and surface damage evaluations also are made by means of SEM. After the realization of bending tests, SEM observation was realized on the fracture surface of the tested specimens with the electron microscope. Creating ideal interfaces for each composition is needed to observe good adhesion of both reinforcements in the rubber-based matrix. Then, density measurements were made by using a pycnometer to determine the volume precisely. Density measurements are variable between 1.85 and 1.92 g/cm^3. Also, mass measurements are recorded by a highly sensitive balance. After curing, hardness measurements on the specimens are performed.

Shore D hardness test measurements on the polished flat surfaces of the specimens are carried out using a Shore D hardness tester. Moreover, dynamic compression tests (i.e., drop-weight tests) are carried out using a drop-weight test device. Impact behavior of the manufactured composites are observed over the test specimens with the help of force–time curves. The closed area of the load-displacement curve is used to determine energy absorption. The energy absorption performance of the specimen is measured by the ratio of the energy absorbed by the specimen to the impact of the energy carried by the impactor.

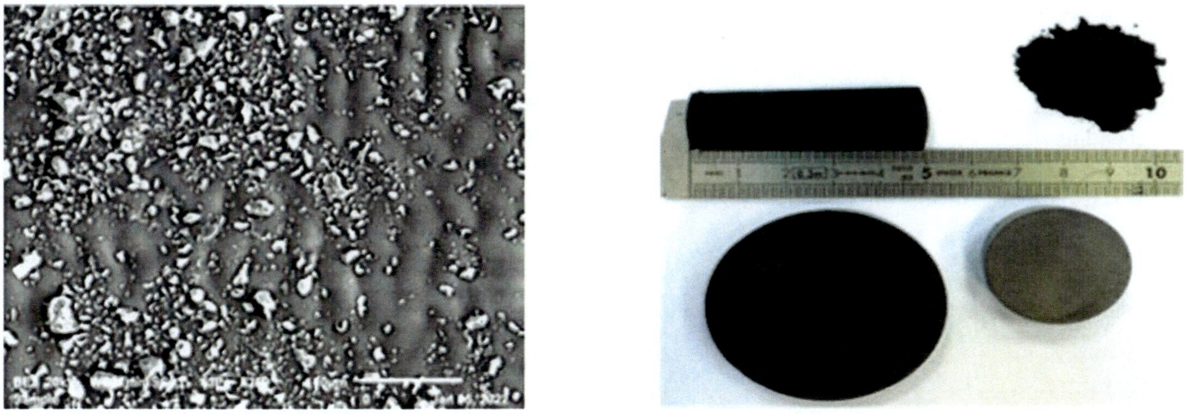

Fig. 2.2 The hBN powder (*left*) as the reinforcement for rubber powder and some samples obtained from the mixed powder

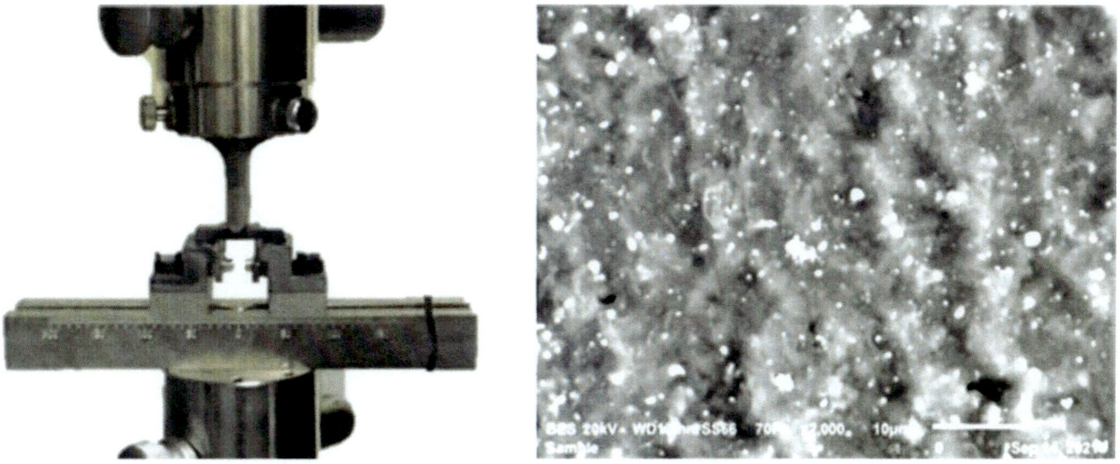

Fig. 2.3 Three-point bending test device adapted on the Zwick test apparatus and microstructure of the sample (BN-I) taken from SEM

2.3 Results and Discussion

Several types of composites were designed starting from fresh scrap rubber composites reinforced with hBN, boron nitrite activated by the electrochemical laboratory in various volume and weight percentages, and the chemical compositions are given in Table 2.1.

All of the micro-hardness measurements have been carried out using Shore D hardness tests to determine the depth of penetration. For Shore D hardness tests, at least five measurements were obtained from the several points on the specimen surface, and the average values are presented in Table 2.1 as well.

The hBN fillers used as reinforcement generally enhanced the surface's hardness. It has been observed that increasing the amount of hBN powders increases imperceptibly the surface hardness. The unexpected results may be because of inhomogeneous distribution of the fillers on the measurement point. For this reason, the results should be accepted as indicator values.

Good distribution of doped nano boron nitride (hBN) in the matrix has shown substantial increments of thermal conductivity and high electrical insulation depending on the quantity in the matrix. Additionally, the thermal analysis indicates that NR/NBN composites have excellent heat-transfer capacity during heating and cooling processes, which suggests the great potential application in thermal conductive and electrical-insulating areas.

The procedure can find multiscale particle-matching ways to achieve the maximum effective thermal conductivity under a given filler load. The optimized effective thermal conductivity obviously can be improved with the increase in the volume fraction of the reinforcement.

In the frame of the present work, some of the thermal conductivity measurements have been carried out in the laboratory scale and the total results are presented in Table 2.2. Also, evolution of these measurements are shown in Fig. 2.4, depending on the amount of hBN reinforcement in the rubber matrix.

General mechanical characterization tests have been carried out. First, static compression test results are presented in Fig. 2.5 with a constant speed of 1 mm/min for the specimens. Then, three-point bending tests have been carried out with a constant speed of 1 mm/min for the specimens; they have been tested for each composition and all the evolution of stress levels, depending on the deformation for each composition, are presented in Fig. 2.6. Comparison of the flexural test results (i.e., ultimate flexural stress, flexural strain and elastic moduli) of the specimens are given in Table 2.3. In the same way, the impact drop-weight tests results are presented in Fig. 2.7. It seems that the light hBN powder has inadequately influenced impact resistance of the specimens.

Table 2.1 Shore D Hardness Measurements of the Composites

Composite design matrix (97wt% Rubber-3wt% epoxy)	wt% of hBN in the matrix	Shore D hardness test
BN1	2	80 ± 3
BN2	4	78 ± 2
BN3	6	80 ± 2
BN4	8	79 ± 2
BN5	16	84 ± 2
BN6	24	85 ± 2
BN7	32	84 ± 2

Table 2.2 Thermal Conductivity Comparison of the Rubber-Based Composites Containing Various Amounts of hBN Powders Added to the Matrix as Reinforcement

Composite	Mass % of hBN	Mass of composite (g)	Specific heat capacity (J/g/°C)	ΔT (°C)	Heat (J)	Δt (s)	Heat transfer (W)	Thermal conductivity (W/m/°C)
BN1	2	25	0.5	55	712.5	1800	0.4514	0.1157
BN2	4	26	0.5	55	712.5	1800	0.46250	0.1156
BN3	6	25	0.5	76	875	1800	0.6680	0.1452
BN4	8	24.7	0.5	75	728.6	1800	0.5515	0.1375
BN5	16	24.6	0.5	82	921.5	1800	0.7564	0.1985
BN6	24	23	0.5	86	1045.4	1800	0.2300	0.2234
BN7	32	24.5	0.5	85	10,294	1800	0.1633	0.2186

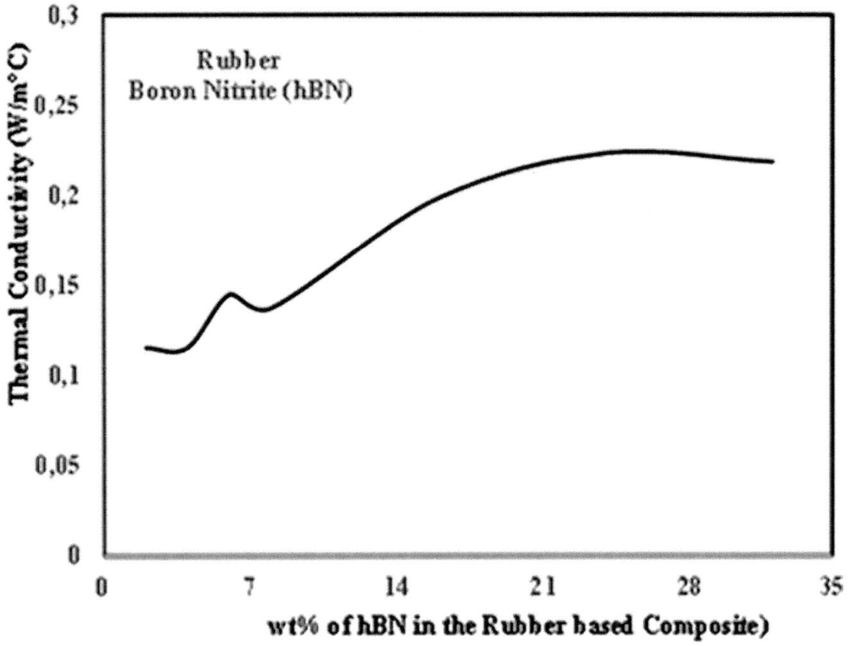

Fig. 2.4 Evolution of the thermal conductivity as a function of hBN reinforcement values added to the rubber matrix

2.4 Conclusions

Recycled natural rubber-based composites reinforced with the doped nano boron nitride (hBN) with the high resolution in thermal conductive and electrical-insulating field were designed and mechanical properties were studied. Seven different compositions in various weight percentages are manufactured to compare the mechanical properties. Good distribution of doped hBN in the matrix has shown substantial increments of thermal conductivity and high electrical insulation depending on the quantity in the matrix. Additionally, the thermal analysis indicates that NR/hBN composites have excellent heat-transfer capacity during heating and cooling processes, which suggests great potential application in thermal conductive and electrical-insulating areas in aeronautical parts.

The procedure can find multiscale particle-matching ways to achieve the maximum effective thermal conductivity under a given filler load. It should be emphasized that the optimized effective thermal conductivity obviously can be improved with the increase in the volume fraction of the reinforcement. This should be carried out with more experimental work to optimize the composite design and to obtain an efficient database for the numerical analyses.

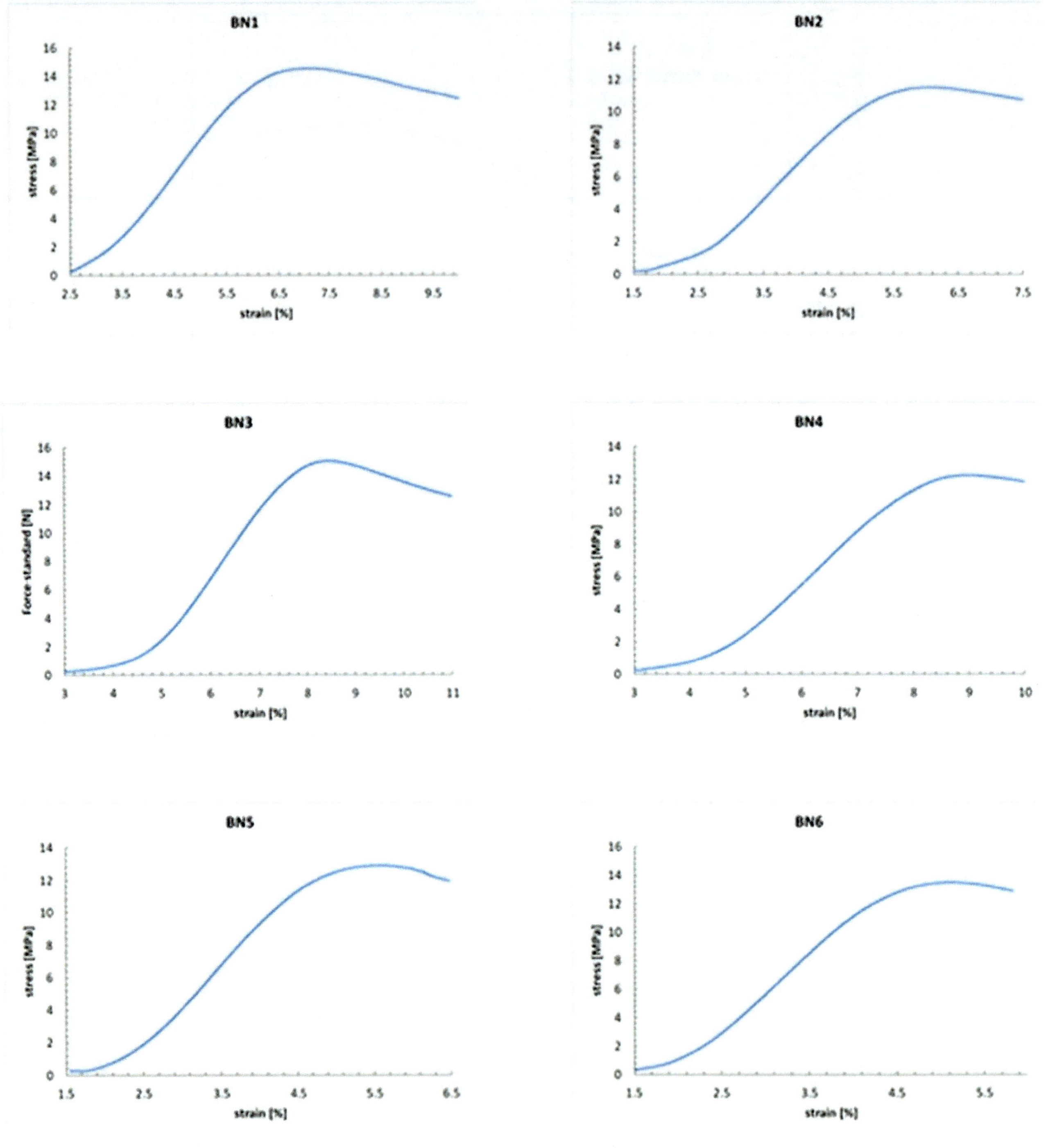

Fig. 2.5 Compression test results for the various compositions of the rubber-based composites containing different amounts of hBN as the reinforcement

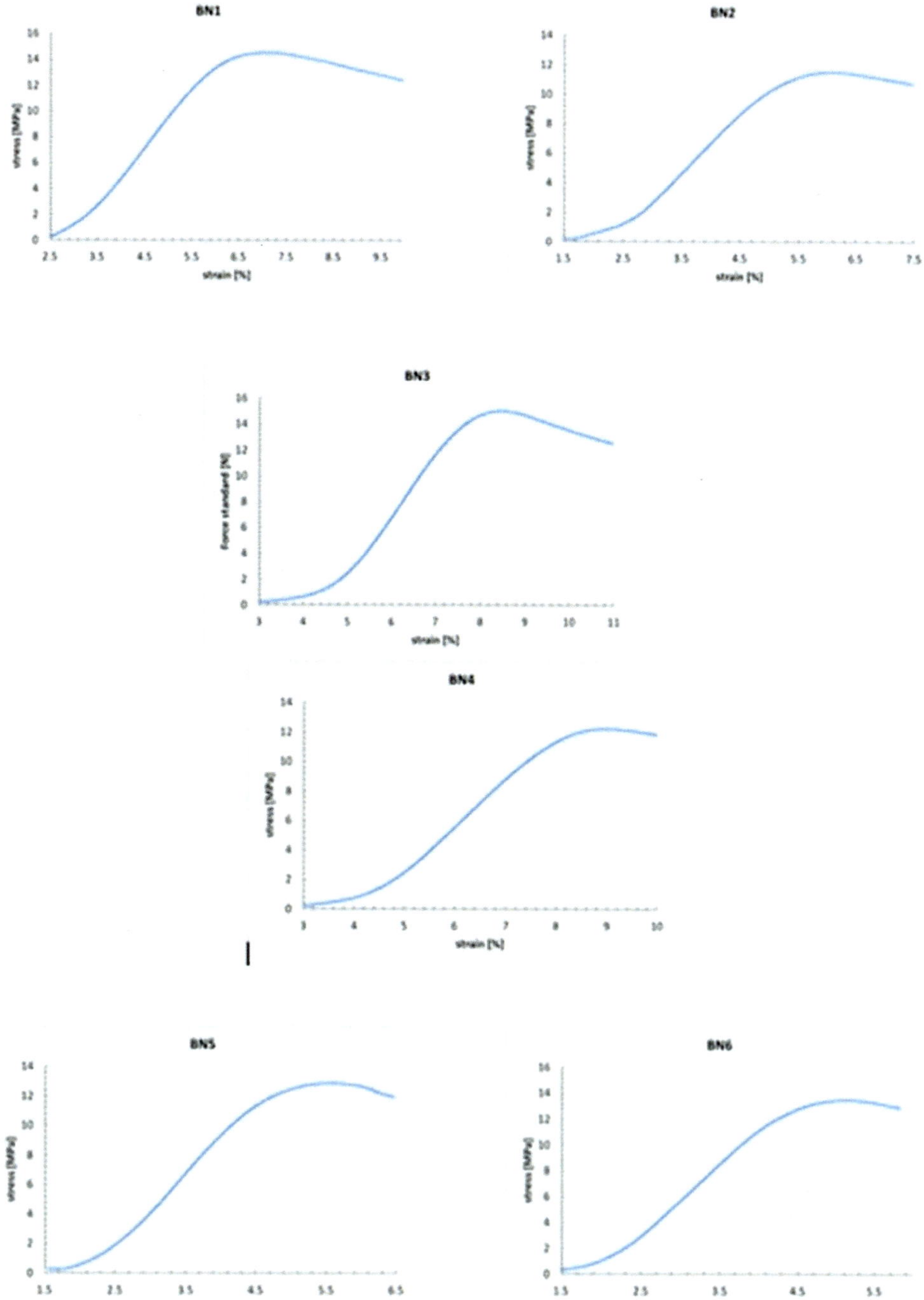

Fig. 2.6 Flexural test results (3 PB) for the various compositions of the rubber-based composites containing different amounts of hBN as the reinforcement

Table 2.3 Comparison of Mechanical Properties of the Specimens Tested

Composition	Weight % of hBN	Ultimate flexural stress, σfB (MPa)	Strain at break, εf (%)	Modulus of elasticity, E (MPa)
BN1	2	11.35	10.92	837
BN2	4	24.95	4.67	857
BN3	6	15.43	4.23	893
BN4	8	9.59	4.44	976
BN5	16	20.64	3.14	1226
BN6	24	18.68	2.64	1670
BN7	32	12.74	3.46	2263

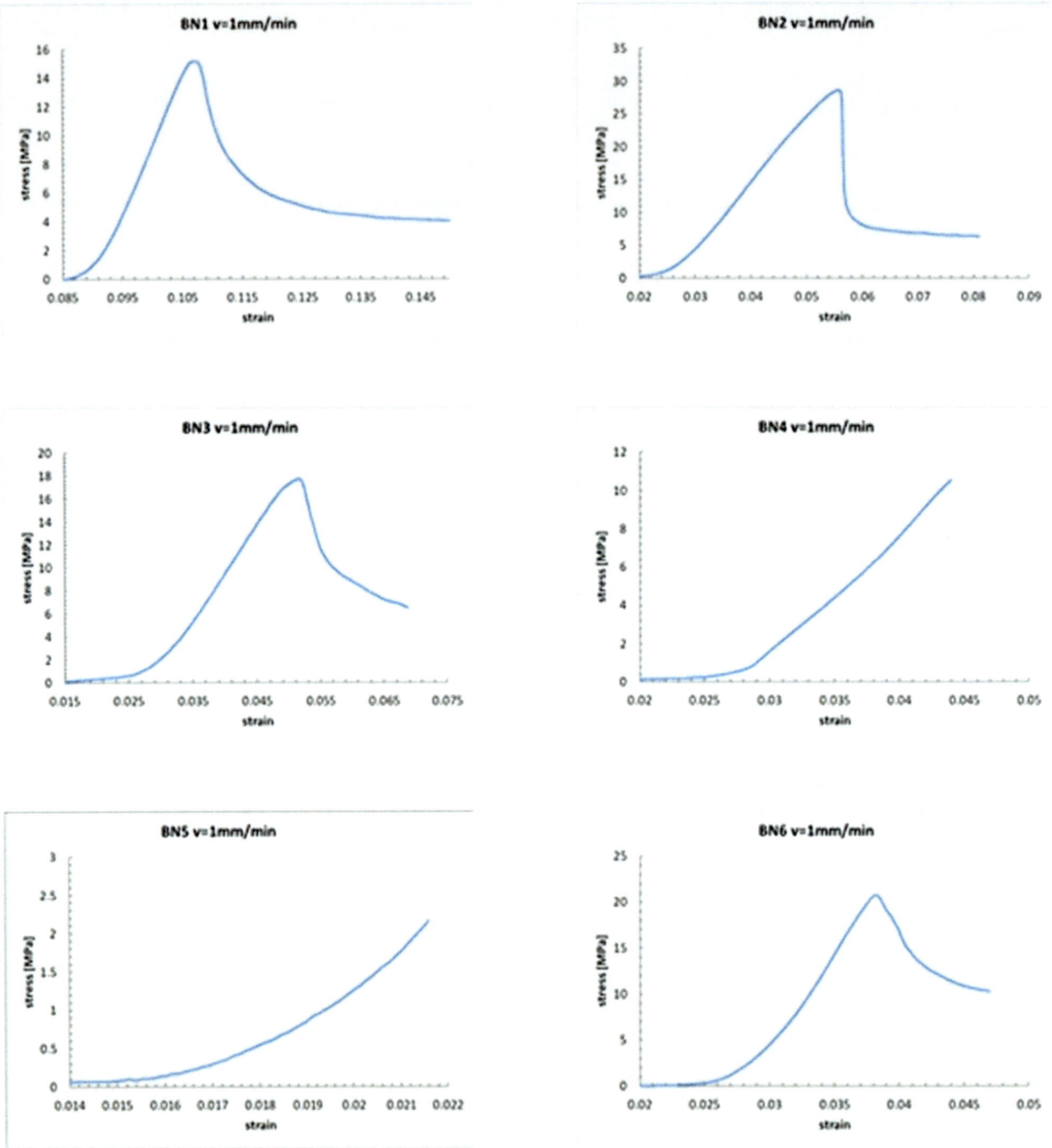

Fig. 2.7 Impact of low velocity (drop-weight) test results measured on the different amounts of the hBN powder reinforcement added to the rubber matrix

Acknowledgements The authors acknowledge Dr. Georges Zambelis, Airbus-Helicopter/Paris, for his support and his valuable discussions for this ongoing research project. Thanks for the technical help of Mr. Christophe Brahim, research assistant at the QUARTZ Lab-ISAE-Supmec-Paris.

References

1. K-Cakir G, Aslan, O., Bayraktar, E.: Numerical modeling of recycled rubber-based composites reinforced with glass fibers at high strain rates, mechanics of composite, hybrid and multifunctional materials, fracture, fatigue, failure and damage. Evolution. **3**, 81–84 (2021)
2. Leung, S.N., Khan, M.O., Chan, E., Naguib, H.E., Dawson, F., Adinkrah, V., et al.: Analytical modeling and characterization of heat transfer in thermally conductive polymer composites filled with spherical particulates. J. Compos. Part B Eng. **45**(1), 43e9 (2013)
3. K-Çakır G, Aslan, Ö., Bayraktar, E.: Toughening mechanism of recycled rubber based composites reinforced with glass fibers + alumina fibers for military applications, mechanics of composite, hybrid and multifunctional materials, fracture, fatigue, failure and damage. Evolution. **3**, 99–110 (2021)
4. Kabakci, G.C., Aslan, O., Bayraktar, E.: Toughening mechanism analysis of recycled rubber-based composites reinforced with glass bubbles, glass fibers and alumina fibers. Polymers. **13**, 4215 (2021). https://doi.org/10.3390/polym13234215
5. Zhao, J.C., Du, F.P., Zhou, X.P., Cui, W., Wang, X.M., Zhu, H., et al.: Thermal conductive and electrical properties of polyurethane/hyperbranched poly(ureaurethane)-grafted multi-walled carbon nanotube composites. J. Compos. Part B Eng. **42**(8), 2111e6 (2011)
6. Zhu, B.L., Wang, J., Zheng, H., Ma, J., Wu, J., Wu, R.: Investigation of thermal conductivity and dielectric properties of LDPE-matrix composites filled with hybrid filler of hollow glass microspheres and nitride particles. J. Compos. Part B Eng. **69**, 496e506 (2015)
7. Mosanenzadeh, S.G., Naguib, H.E.: Effect of filler arrangement and networking of hexagonal boron nitride on the conductivity of new thermal management polymeric composites. Compos. Part B. **85**, 24–30 (2016) ISSN 1359–8368
8. Kawaguchi, M., Kuroda, S., Muramatsu, Y.: Electronic structure and intercalation chemistry of graphite-like layered material with a composition of BC6N. J. Phys. Chem. Solids. **69**(5e6), 1171 (2008)

Chapter 3
Impact Behaviour of Recycled Rubber-Based Composites Reinforced with Glass Bubbles and Alumina Fibers (γ-Al_2O_3)

Gamze Çakır Kabakcı, Özgür Aslan, and Emin Bayraktar

Abstract In this research, recycled rubber-based composites are considered with glass bubble (GB) and fine gamma alumina fiber (γ-Al_2O_3) reinforcements. The effect of the reinforcements with matrix, fracture characteristics of the composite are studied by impact-loading tests (i.e., drop-weight tests). These test results are simulated by finite element analysis (FEM) and the results are compared with the experimental results. Microstructural and fracture surface analysis are carried out by means of scanning electron microscopy (SEM). Mechanical test results show that the reinforcement with glass bubbles and aluminum oxide ceramic fibers generally increase the damping capacity and fracture toughness of the composites.

Keywords Recycled rubber · Glass bubble · Alumina fiber · Scanning electron microscopy

3.1 Introduction

Low-cost devulcanized recycled rubber-based composites are reinforced with glass bubble (GB) microspheres and fine gamma alumina fibers (γ-Al_2O_3) to provide multifunctionality to the composites. In this study, glass bubbles are preferred as one of major reinforcements because they lead to perfect weight reduction, performance, processing and dimensional stability characteristics when used with other reinforcing fillers [1]. Also, glass bubbles have a thin-walled hollow spherical shape, soda-lime-borosilicate glass composition and high strength to density ratio. The hollow glass bubbles have a density of 0.65 g/cm^3, average diameter of 30 μm, and modulus of elasticity of 3546 MPa. A perfect spherical shape makes them efficient volume fillers because they can sustain high volumetric impact loads without increasing the viscosity to unacceptable levels for further polymer operations. Then, fine gamma alumina fibers with the density of 0.65 g/cm^3, diameter of 5–10 μm and the modulus of elasticity of 340 GPa are preferred because they generally improve fracture toughness, resin stiffness and wear resistance and have high compressive strength and shear properties [2].

Alumina also has good thermal conductivity and it is inexpensive, nontoxic and highly abrasive [3–5]. Figure 3.1 shows scanning electron microscopy (SEM) images of the reinforcements used in this study. In addition, recycled rubber is supplied by the sports equipment manufacturers as fresh scrap collected directly from the production line. The average diameter of the rubber particles is 50 μm, the modulus of elasticity is 10 MPa, the elongation at break is 80–100%, the hardness is 37–40 Shore A and the density is 1 g/cm^3. Certain steps to obtain rubber powder from fresh scrap thin rubber is presented in Fig. 3.2.

Furthermore, the level of cross-linking has a significant effect on the mechanical properties [6]. The toughness improvement in hybrid rubber-based composites can increase the plastic zone in the structure and this leads the rubber-based composites to dissipate additional fracture energy. For this reason, the mixture of composite material is blended with epoxy resin (3 wt.%).

G. Çakır Kabakcı
Department of Mechanical Engineering, Atilim University, Ankara, Turkey

School of Mechanical and Manufacturing Engineering, ISAE-Supmeca-Paris, Paris, France
e-mail: gamze.cakir@atilim.edu.tr

Ö. Aslan
Department of Mechanical Engineering, Atilim University, Ankara, Turkey
e-mail: ozgur.aslan@atilim.edu.tr

E. Bayraktar (✉)
School of Mechanical and Manufacturing Engineering, ISAE-Supmeca-Paris, Paris, France
e-mail: emin.bayraktar@isae-supmeca.fr

V. Chalivendra, F. Gardea (eds.), *Mechanics of Composite, Hybrid & Multi-functional Materials, Volume 5*, Conference Proceedings of the Society for Experimental Mechanics Series, https://doi.org/10.1007/978-3-031-17445-2_3

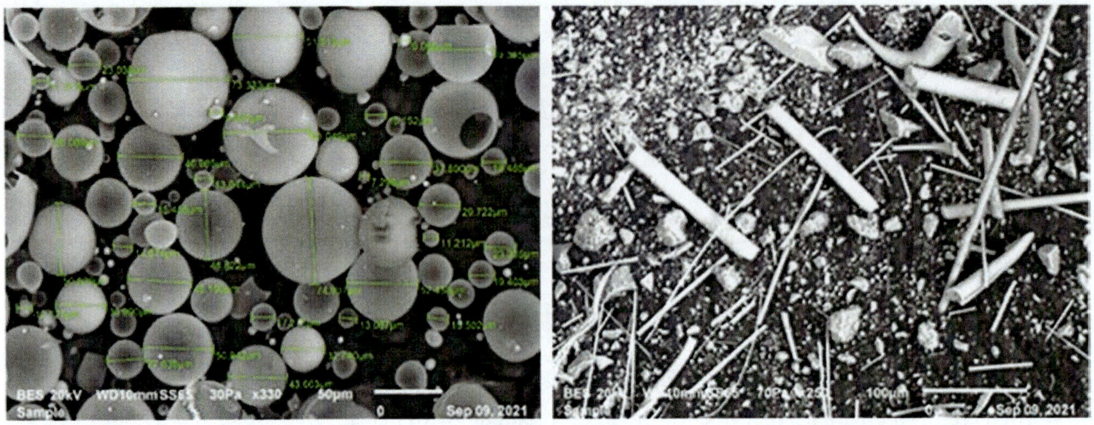

(a) (b)

Fig. 3.1 SEM images of (**a**) glass bubble microspheres and (**b**) alumina fibers (γ-Al_2O_3)

Fig. 3.2 Steps from fresh scrap thin rubber to rubber powder

3.2 Material Processing and Characterization Tests

For manufacturing of recycled rubber-based composite, a good quality of bonding between matrix and reinforcements is required; however, recycled rubbers do not carry any free chains to form new bonds with epoxy resin because of the previous vulcanization process. To improve the interface quality between epoxy and rubber, fresh scrap rubber should be devulcanized. During the devulcanization process, Sulphur links are attempted to be broken and new links are generated, then the flowing capacity and interaction of recycled rubbers with other substances are increased. After determination of the reinforcements with matrix in wt%, a special process is applied to complete the manufacturing of the composites successfully.

This process involves Silanization of the recycled rubber and devulcanization before blending it with epoxy resin and reinforcement. Chemical treatment followed devulcanized recycled rubber is mixed with epoxy resin to obtain strong chemical bonding diffusion reinforced with various reinforcements. After blending of the mixture, they are milled during 4 h by adding "Zn-Stearate" as a lubricant for homogenous distribution.

For manufacturing of the composites, hot compaction is carried out at 230 °C during 20 min under the pressure of 6 tons after that the specimens produced have a diameter of 50 mm; a similar procedure is shown in Fig. 3.3. After manufacturing of the composites, specimens are cut into rectangular shape by waterjet or metallographic cut-off machine for flexural and impact load testing. Figure 3.4 shows the rubber powder and the cut specimen. A three-point bending (3 PB) test is conducted, as seen in Fig. 3.5. Deflection of the specimen is measured by the crosshead position. Flexural strength, modulus of elasticity in bending and strain are obtained from the test results.

Microstructure and surface damage evaluations also are made by means of SEM. After the realization of bending tests, SEM observation is realized on fracture surface of the tested specimens with an electron microscope. Creating an ideal

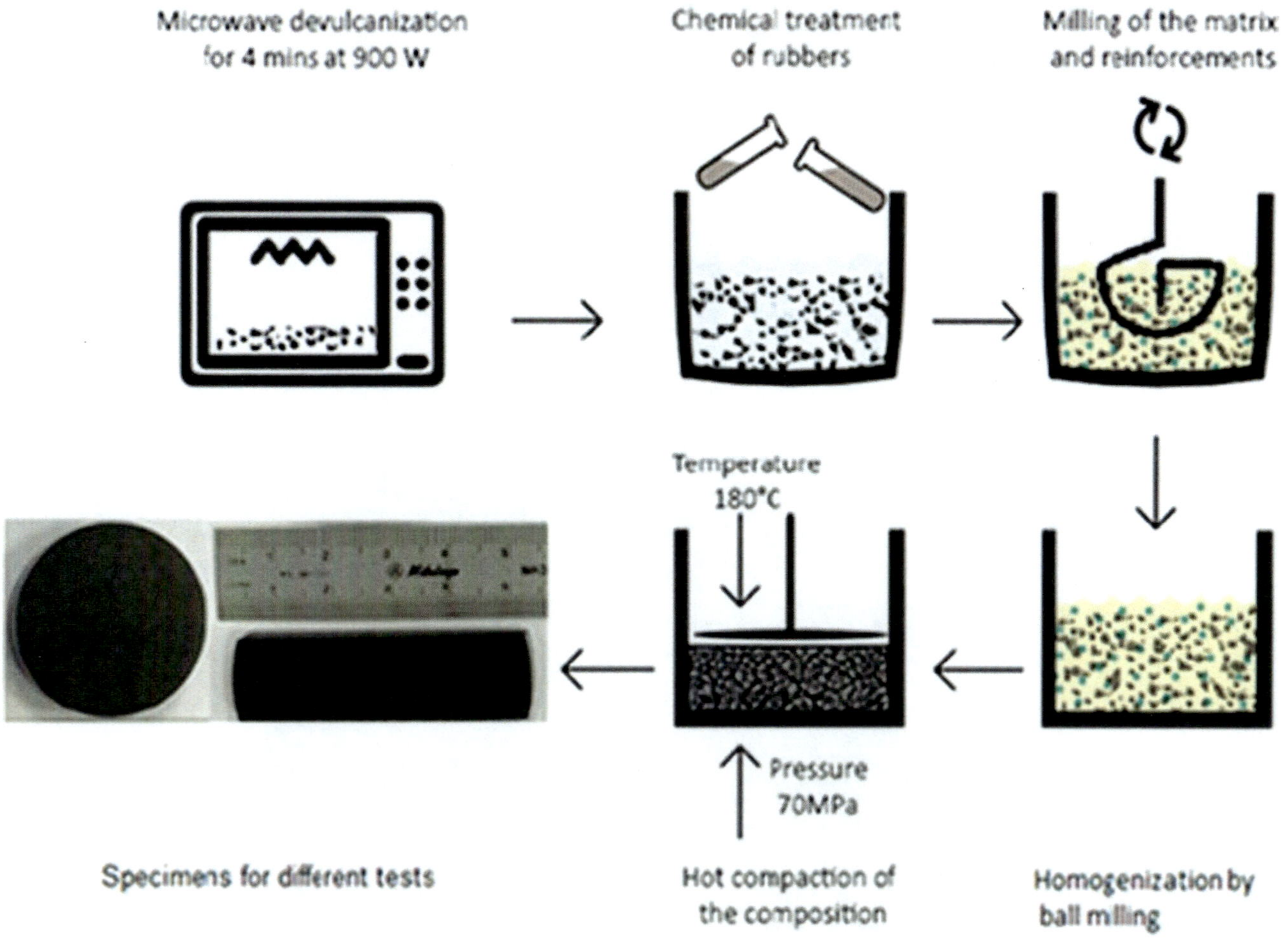

Fig. 3.3 Manufacturing steps of recycled rubber-based composites [5]

Fig. 3.4 The rubber powder and the cut specimen

interface for each composition is needed to observe good adhesion of both of the reinforcements in the rubber-based matrix. Debonding of nanoparticles with consequent void growth can be identified in the structure as the improved toughening mechanisms. Then, density measurements are made by using a pycnometer to determine the volume precisely. Also, mass measurements are recorded by a highly sensitive balance. After curing, hardness measurements on the specimens are performed.

Hardness test measurements on the polished flat surfaces of the specimens are carried out using a Shore D hardness tester. Moreover, dynamic compression tests (drop-weight tests) are carried out using a drop-weight test device. Impact behaviour

Fig. 3.5 Three-point bending test

of the manufactured composites are observed over the test specimens with the help of force–time curves. The closed area of the load-displacement curve is used to determine energy absorption. The energy absorption performance of the specimen is measured by the ratio of the energy absorbed by the specimen to the impact energy carried by the impactor.

3.3 Results and Discussion

Several recycled fresh scrap rubber composites have been designed and manufactured with glass bubble (GB) and fine gamma alumina (γ-Al_2O_3) fiber (AF) reinforcements in various volumes and weight percentages, and chemical compositions are given in the Table 3.1.

All the microhardness measurements have been carried out by using Shore D hardness tests to determine the depth of penetration. For the Shore D hardness test, at least five measurements were obtained from the various points on the specimen surface, and the average values are presented in Table 3.2. The fillers used as reinforcement generally enhance the surface hardness. It was observed that increasing the amount of alumina fibers weakly increased the surface hardness, whereas the amount of the glass bubbles had a minor effect on the surface hardness. The unexpected results may be because of inhomogeneous distribution of the fillers on the measurement point. For this reason, the results should be accepted as an indicator value. More data is needed to optimize these results.

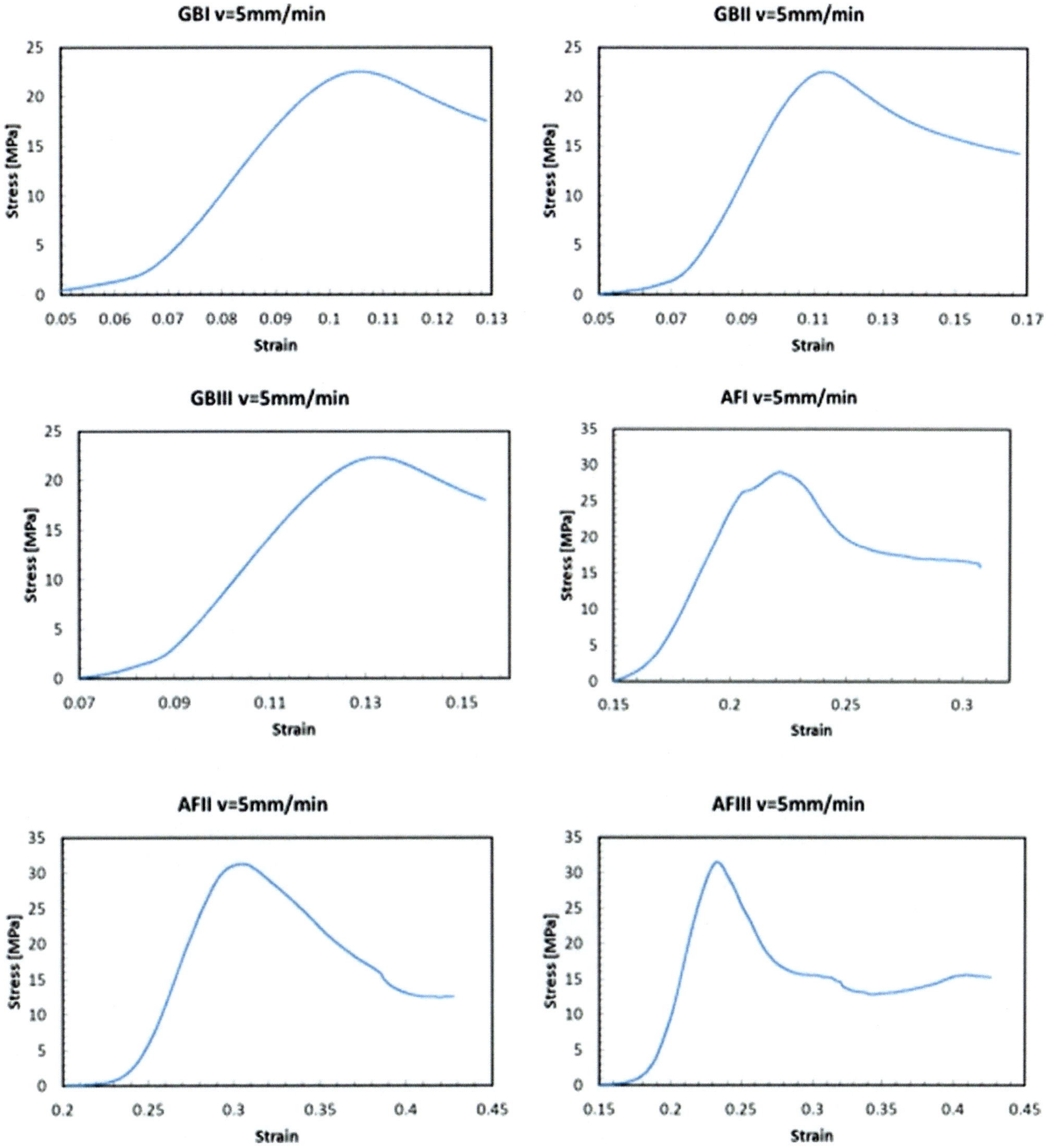

Fig. 3.6 Three-point bending test results for the various compositions

Table 3.1 Chemical Compositions of the Composites

Composite name	Matrix (rubber/epoxy)	Glass bubble (vol %)	γ-Al_2O_3 fiber (wt%)
GBI	95 (97/3)	5	
GBII	90 (97/3)	10	
GBIII	85 (97/3)	15	
AFI	95 (97/3)		5
AFII	90 (97/3)		10
AFIII	85 (97/3)		15

Table 3.2 Shore D Hardness Measurements of the Composites

Composition γ-Al_2O_3 fiber	Hardness (Shore D)	Composition glass bubble	Hardness (Shore D)
AF1	84 ± 5	GB1	76 ± 2
AF2	87 ± 2	GB2	79 ± 2
AF3	89 ± 5	GB3	79 ± 5

Table 3.3 Comparison of Mechanical Properties of the Specimens

Composition name	Density of the composite (g/cm^3)	Ultimate flexural stress, σ_{fB} (MPa)	Strain at break, ε_f (%)	Modulus of elasticity, E (MPa)
GBI	1.418	22.54	10.56	381.59
GBII	1.379	22.53	11.32	436.4
GBIII	1.337	22.30	13.24	488.61
AFI	1.465	29.15	20.65	709.92
AFII	1.468	31.27	30.50	922.49
AFIII	1.472	31.51	23.30	1138.45

Table 3.4 Average Results for the Energy Absorption Performance

Composite name	Impact energy, K (J)	Energy absorbed, E (J)	Energy absorption (%)
Pure rubber	9.728	4.250	43.69
GBII	9.728	5.251	53.98
GBIII	9.728	5.507	56.61
AFII	9.728	6.068	62.38
AFIII	9.728	6.236	64.11

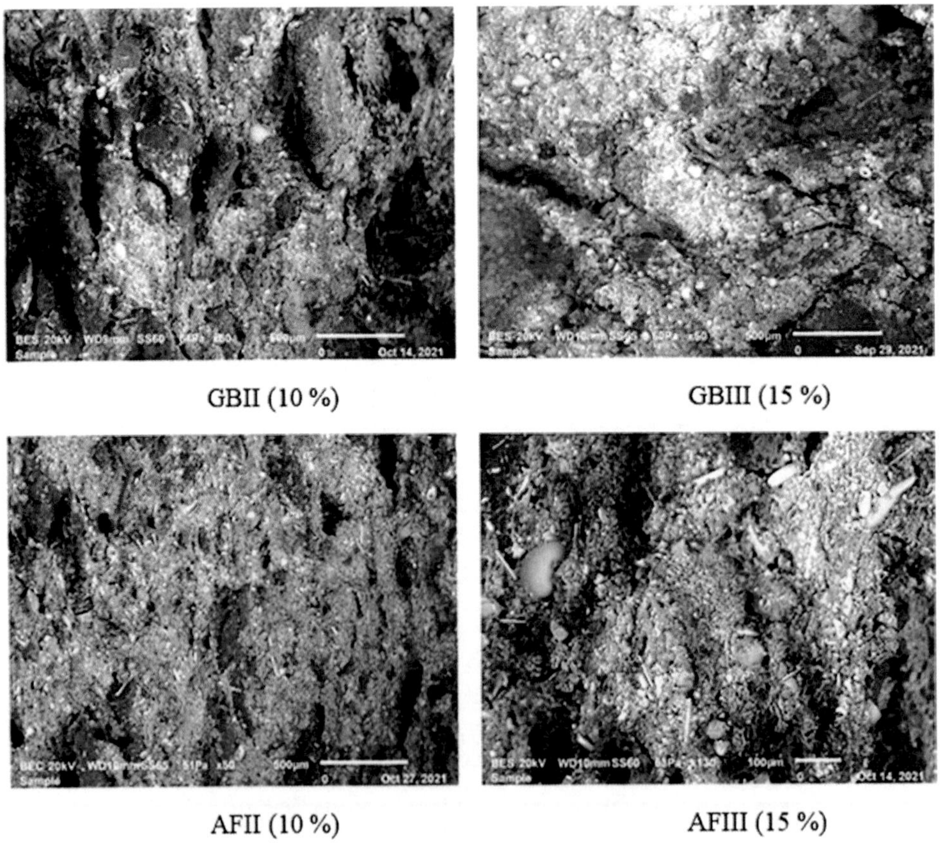

Fig. 3.7 Microstructure and fracture surface analyses for the composites

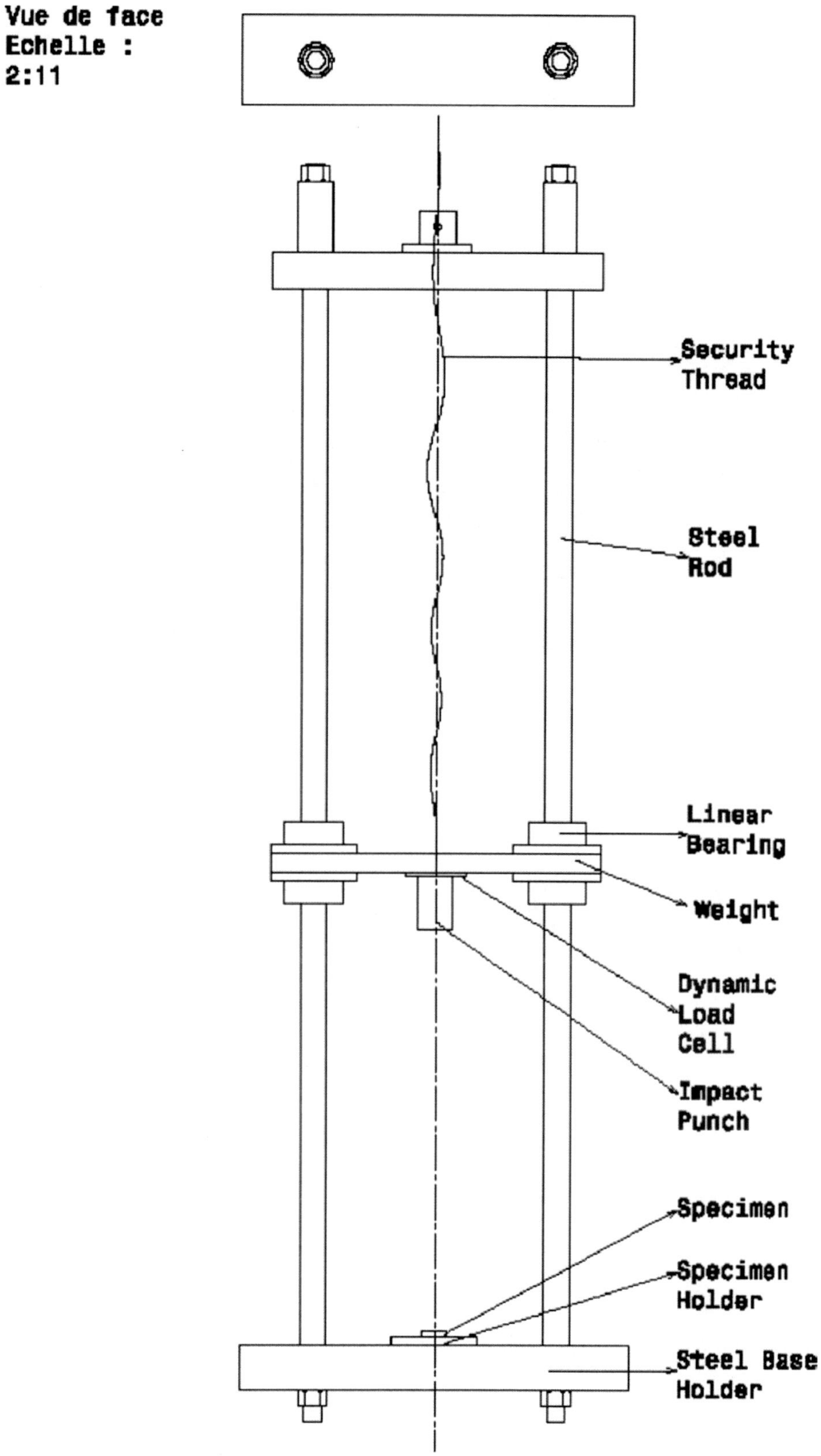

Fig. 3.8 Schematic drawing of drop-weight test

Fig. 3.9 Drop-weight test apparatus

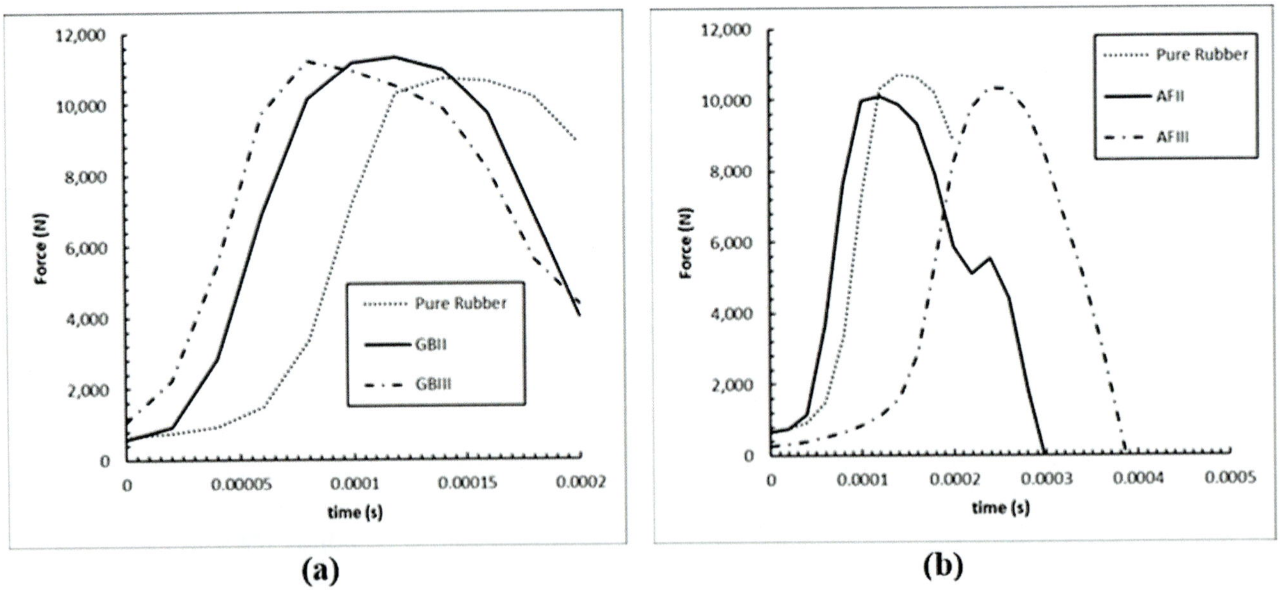

Fig. 3.10 Force–time curve from drop-weight test for composites (a) GBII and GBIII and (b) composites AFII and AFIII

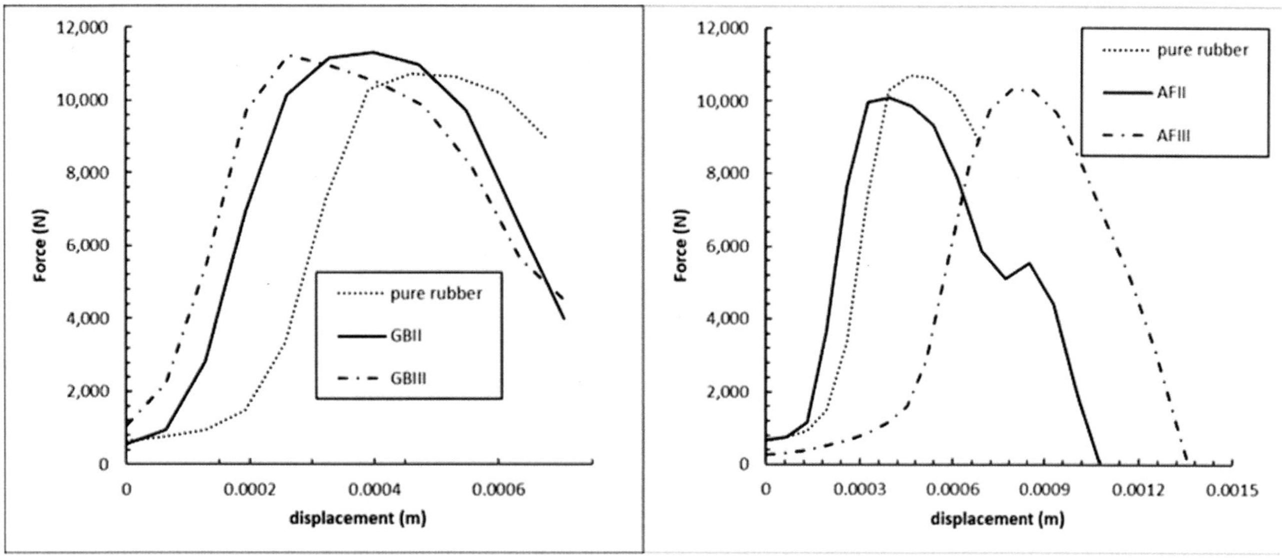

(a) (b)

Fig. 3.11 Force displacement curve from drop-weight test for composites (**a**) GBII and GBIII and (**b**) composites AFII and AFIII

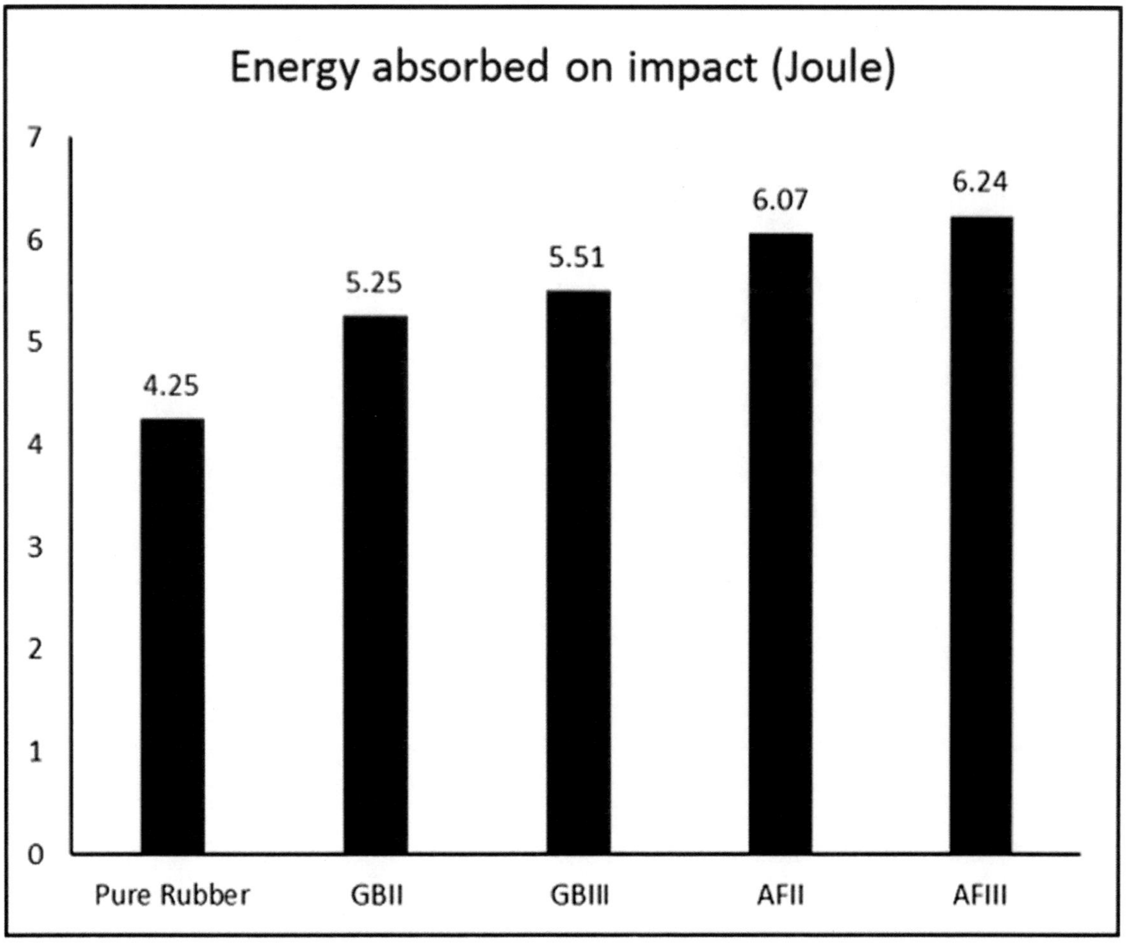

Fig. 3.12 Energy absorbed on impact

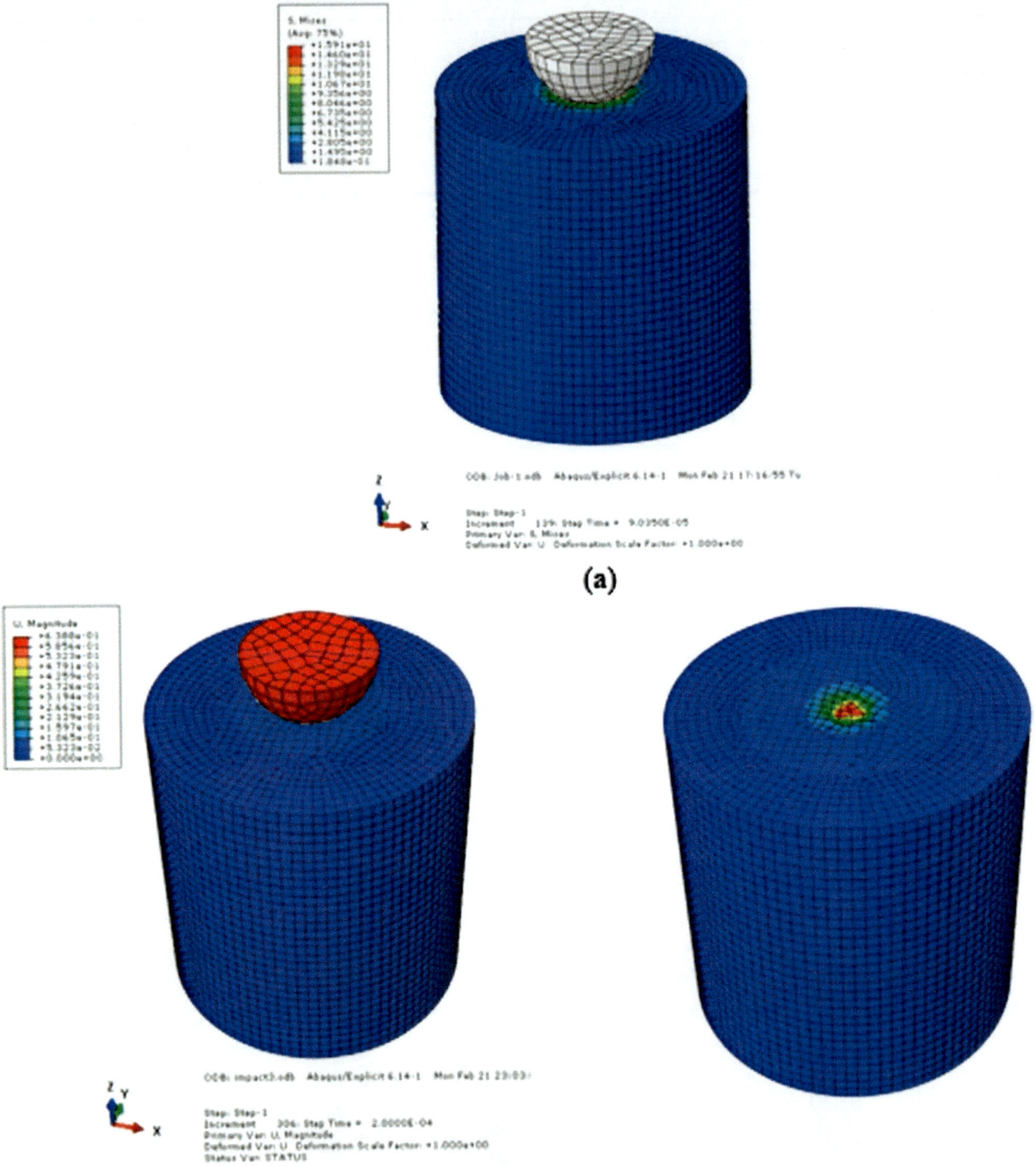

Fig. 3.13 Numerical simulation of drop-weight test for the composite GBIII (**a**) von Mises stress at the time the maximum force is applied and (**b**) the displacement at $t = 0.0002$ s

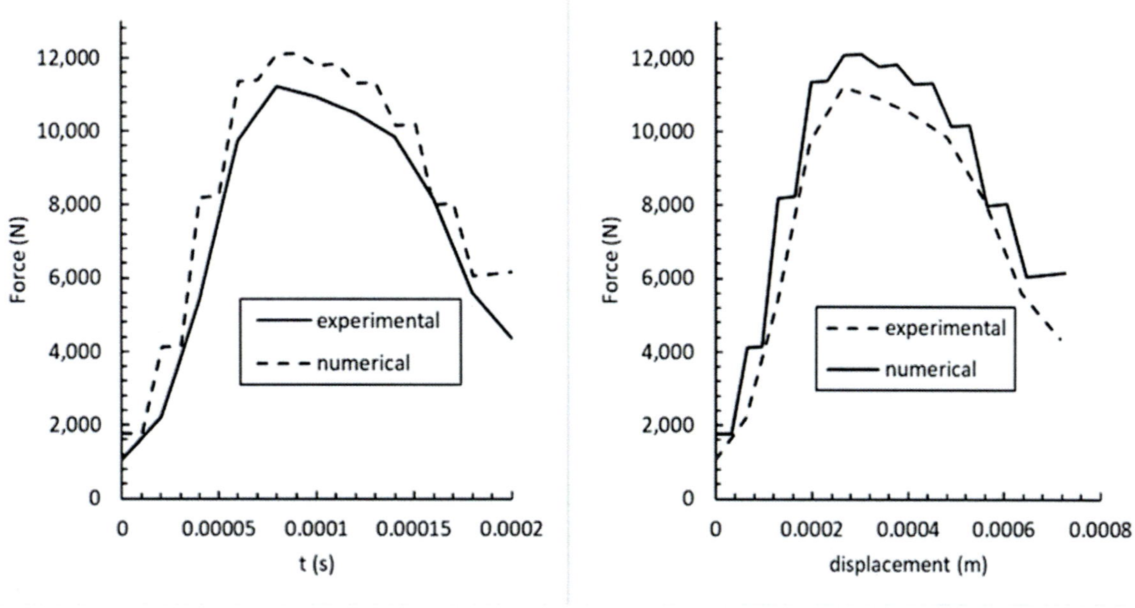

Fig. 3.14 Comparison of the force–time and force–displacement curves of the experimental study and numerical analysis for impact test of composite GBIII

General mechanical characterization tests have been carried out. Final rectangular shape specimens have been cut for flexural tests. 3 PB tests have been carried out with a constant speed of 5 mm/min for the specimens, and they have been tested for each composition and evolution of stress levels depending on the deformation for each composition; results are presented in Fig. 3.6.

The rubber-based composites reinforced with glass bubbles and alumina fibers with a certain ratio have shown improvements in the flexural stress and the modulus of elasticity of the matrix. Yet, the density of the composite decreases by increasing the ratio of glass bubble microspheres. The comparison of the flexural test results and the density of the composites are presented in Table 3.3.

After the realization of bending tests, fracture surface analyses were conducted by using SEM on the damaged specimens. Figure 3.7 gives the details on the microstructure evolution with various reinforcements that a homogenous distribution of the reinforcement in the matrix and a strong cohesion between the reinforcements and matrix can be observed.

Drop-weight tests were conducted and impact behaviour of the composites were observed by force–time curves. The drop-weight test apparatus consists of a typical block weighting 1900 g containing a load cell inside of hemispherical punch with a diameter of 12.5 mm. All the impact tests were carried out from 0.9 m. The drop mass is 1.9 kg and the speed at impact is $3.2\frac{\mathrm{m}}{\mathrm{s}}$, taking into account the friction for the tests. The impact energy is calculated as 9.728 J. The schematic drawing and drop-weight test apparatus used in this study are presented in Figs. 3.8 and 3.9, respectively.

Force–time curves are presented for compositions GBII, GBIII and AFII, AFIII comparing the response of the pure rubber matrix in Fig. 3.10a, b, respectively. The force displacement curves are presented for compositions GBII, GBIII and AFII, AFIII in Fig. 3.11a, b, respectively. As seen in Fig. 3.11, the region beneath the load-displacement curve graphically represents the absorbed energy. Energy absorbed on impact and energy absorption performance are presented in Table 3.4 and Fig. 3.12.

The finite element model is developed using the software Abaqus 6.14. 29178; quadrilateral finite elements are generated based on the physical configuration of the experimental drop-weight test. The geometry comprises the specimen and the impact punch. The numerical analysis for the composite GBIII is presented in Fig. 3.13. The material model is elastoplastic and isotropic with a strain rate of 109.5 s^{-1}. The impact velocity is taken as 3.2 m/s as in the experimental test.

Figure 3.14 shows the comparison of the force–time and force–displacement responses of experimental and numerical analysis for the impact test of composite GBIII. The impact energy is 9.728 J, and the energy absorbed on impact is obtained as 6.29 J from numerical analysis. The numerical analysis can be used as a preliminary evaluation to estimate how much energy is necessary to cause complete failure of the specimen.

3.4 Conclusions

Recycled rubber-based composites are considered with glass bubble (GB) and fine gamma alumina fiber (γ-Al_2O_3) reinforcements. The effect of the reinforcements with matrix, fracture characteristics of the composite were studied by impact-loading tests. These test results were simulated by finite element analysis (FEM) and compared with the experimental results.

Microstructural and fracture surfaces analysis were carried out by means of scanning electron microscopy (SEM). Mechanical test results showed that the reinforcement with glass bubbles and aluminum oxide ceramic fibers generally increased the damping capacity and fracture toughness of the composites.

Dynamic compression tests were carried out and the impact behaviour of the manufactured composites were observed over the test specimens with the help of force–time curves. Force–time curves and force–displacement curves were plotted for various compositions comparing the response of the pure rubber matrix.

The energy absorption performance of the specimen was measured by the ratio of the energy absorbed by the specimen to the impact energy carried by the impactor. It can be stated that the reinforcements increased the energy absorption performance of the rubber matrix. The numerical analysis carried out for the elastoplastic and isotropic material model for the drop-weight test was presented.

The results were compared with the experimental test. This numerical analysis can be used as a preliminary evaluation to estimate how much energy is necessary to cause complete failure of the specimen. With this energy value, it is possible to develop experimental tests using the minimum requirements of the test machine and to avoid additional wasted energy being absorbed by the composite.

References

1. Zhang, Z., Lei, H.: Preparation of α-alumina/polymethacrylic acid composite abrasive and its CMP performance on glass substrate. Microelectron. Eng. **85**(4), 714–720 (2008)
2. Omairey, S.L., Dunning, P.D., Sriramula, S.: Development of an ABAQUS plugin tool for periodic RVE homogenisation. Eng. Comput. **35**, 567–577 (2019)
3. Li, G., Sharifpour, F., Bahmani, A., Montesano, J.: A new approach to rapidly generate random periodic representative volume elements for microstructural assessment of high-volume fraction composites. Mater. Des. **150**, 124–138 (2018) ISSN 0264-1275
4. Yalcin B, Amos SE, Williams MJ, Gunes IS, Ista TK *3M™ Glass Bubbles iM16K for Reinforced Thermoplastics*, 2016. www.3M.com/glassbubbles
5. Williams, J.G.: *Fracture Mechanics of Polymers*. Horwood (1984)
6. Amirshaghaghi, Kokabi, M.: Tailoring size of α-Al^2O^3 nanopowders via polymeric gel-net method. Iran. Polym. J. **19**, 615–624 (2010)
7. Cakir, G.K., Aslan, Ö., Bayraktar, E.: Numerical modeling of recycled rubber based composites reinforced with glass fibers at high strain rates, mechanics of composite, hybrid and multifunctional materials, fracture, fatigue, failure and damage. Evolution. **3**, 81–84 (2021)
8. Çakır, G.K., Aslan, Ö., Bayraktar, E.: Toughening mechanism of recycled rubber based composites reinforced with glass fibers + alumina fibers for military applications, mechanics of composite, hybrid and multifunctional materials, fracture, fatigue, failure and damage. Evolution. **3**, 99–110 (2021)
9. Kabakci, G.C., Aslan, O., Bayraktar, E.: Toughening mechanism analysis of recycled rubber-based composites reinforced with glass bubbles, glass fibers and alumina fibers. Polymers. **13**, 4215 (2021). https://doi.org/10.3390/polym13234215

Check for updates

Chapter 4
Toughening Mechanism of Silicon Whiskers and Alumina Fibres (γ-Al_2O_3) Reinforced Ni-Al-Cu Matrix Composites Through "Sintering + Forging"

Ibrahim Miskioglu, Georges Zambelis, Fabio Gatamorta, Ozgur Aslan, and Emin Bayraktar

Abstract In this study, the microstructural formation and static/dynamic compression behaviour of recycled Ni-Al-Cu matrix hybrid composites reinforced with silicon whiskers and alumina (Al_2O_3) fibres will be studied. It is intended to be an alternative to traditional alloys/composites used in the aeronautical industry. These composites generally are produced using by combined "sintering + forging" processes. The static and dynamic properties will be evaluated in detail, considering the relevant scanning electron microscopy (SEM) microstructures (including the distribution of reinforcement elements).

Keywords Recycled composites · SiCws · γ-Al_2O_3 · SEM · Static and dynamic tests

4.1 Introduction

Nickel aluminide (Ni-Al) intermetallic-based composites are considered as a better alternative to Ni-based super alloys because of their low density with great oxidation resistance and thermal conductivity for aeronautical engineering applications. Ni-Al metal matrix composites provide high toughness and strength at high temperatures that can match the super alloys used in industry while keeping the weight of the aircraft components as small as possible.

To achieve high strength responses at operating temperatures of aeronautical applications (600–850 °C), intermetallic compounds (e.g., SiCws, γ-Al_2O_3, Nb_2Al, Zr_2O_3 and so on) are preferred as reinforcement materials [1–10]. For the past decade, the focus of several studies has been on the improvement of metal matrix composites by using particle reinforcements on micro and nano scales.

Development of these hybrid composites opens new possibilities in the automotive industry as well as aeronautical and aerospace applications in terms of material characteristics because the mechanical properties of metal matrix-hybrid composites can be modified according to the specific needs of the application area [3–14].

Basically, addition of nanoparticles along with fine ceramic particles improves the mechanical properties of the composites much more efficiently; as a result, it is possible to obtain the required strength and toughness along with the ductility of the composites [8, 9–20]. In fact, a combined powder metallurgy route can be very easily applied as an economic manufacturing process; this means that sintering + forging of powders can be obtained easily from scrap metallic parts with a healthy

I. Miskioglu
ME-EM Department, Michigan Technological University, Houghton, MI, USA
e-mail: imiski@mtu.edu

G. Zambelis
Airbus-HelicoptersComposites Research and Development Centre, Paris, France
e-mail: georges.zambelis@airbus.com

F. Gatamorta
UNICMP-FEM, University of Campinas, Campinas, SP, Brazil
e-mail: fabiog@fem.unicamp.br

O. Aslan
Department of Mechanical Engineering, Atilim University, Ankara, Turkey
e-mail: ozgur.aslan@atilim.edu.tr

E. Bayraktar (✉)
School of Mechanical and Manufacturing Engineering, ISAE-SUPMECA-Paris, Paris, France
e-mail: emin.bayraktar@isae-supmeca.fr

V. Chalivendra, F. Gardea (eds.), *Mechanics of Composite, Hybrid & Multi-functional Materials, Volume 5*, Conference Proceedings of the Society for Experimental Mechanics Series, https://doi.org/10.1007/978-3-031-17445-2_4

microstructure. In other words, the micro-defects of the scrap parts are practically eliminated; therefore, it is possible to obtain the lightweight metal-matrix composite approaching the theoretical densities of the composites [10–16].

This work sought to study the compression behaviour of recycled Ni-Al matrix-hybrid composites reinforced with SiCws, γ-Al_2O_3 under static and dynamic conditions for various particle reinforcement compositions. Mechanical properties, microstructural and fracture surface analyses have been carried out by means of scanning electron microscopy (SEM).

4.2 Experimental Conditions

In the frame of the common research project, five different composites were designed from fresh scrap + gas atomized Ni-60wt.%Al intermetallic matrix with the reinforcement of pure electrolytic copper and silicon whiskers and alumina (Al_2O_3) fibres together. All the reinforcement powders of purity were at the level of 99.5%.

Five types of composites were processed by using a powder metallurgy-combined method: sintering + forging. These composites were sintered at the same temperature (650 °C) followed by forging. Also certain samples of the first group were only sintered to compare with the combined method. Accordingly, mechanical properties of the sintered forging composites were considerably improved.

In Table 4.1, the general compositions of the three composites are presented with the major and minor reinforcements. These compositions actually are under development to optimize the interface matrix-reinforcements for the common research project that is ongoing.

The microstructure analyses have been carried out by SEM with "EDS" chemical analyses. The density measurements of the specimens were carried out by using the Archimedes method. These values change between 5.95 and 6.75 g/cm^3. All the mechanical tests have been carried out on a Zwick mechanical test device installed at the laboratory of ISAE-SUPMECA-Paris, France.

Three prepared cylindrical specimens (H/D≥1.5) were tested under quasi-static and cyclic compression conditions according to the DIN 50106 norm by using a constant test speed as 1 and 1,5 mm/min. Finally, low-velocity impact (drop-weight) tests were carried out with a drop tower to evaluate the response of the composites to dynamic loading. Three cylindrical specimens (H/D≥1.5) were used under low-velocity/drop-weight test conditions.

Microhardness ($HV_{0.1}$) tests were conducted on the polished and etched specimens. The mean microhardness values are presented in Table 4.2 with ±15–20% accuracy for the two manufacturing processing techniques, respectively.

4.3 Results and Discussions

Microstructure analyses starting from the analyses of Ni-Al intermetallic and powder evaluation of the Ni-Al-Cu-based composite are shown in Fig. 4.1. "EDS" chemical analyses also are presented in this figure.

As usual, intermetallic phases occurred with a strong chemical fusion during the sintering. Addition of pure electrolytic copper powder improved the strong chemical-bonding diffusion between matrix and the reinforcements; essentially, ceramic particles were firmly attached in the matrix without a decohesion with the matrix. More detailed analyses should be required in the case of the high number of ceramic particles in the matrix. These results are limited under the laboratory conditions. Static compression test results are presented schematically in Fig. 4.2.

Figure 4.3 shows typical microstructural evaluation with elementary mapping analyses for only the composite AL-3. It seems that the distribution of each element in the structure homogenously without agglomeration. This has a real effect on the mechanical behaviour and resistance of the composites under static and cyclic solicitations.

Table 4.1 Compositions of the Specimens (wt.%)

Specimen name	Ni-Al Fresh scrap	AA7075 Fresh scrap	Electrolitic copper Cu	γ-Al_2O_3 Fibre	SiC whiskers	hBN 0,5	GNP	Zn St
AL-I	B	30	30	3	0,5	0,5	0,5	2
AL-II	B	25	30	3	0,5	0,5	0,5	2
AL-III	B	20	30	3	0,5	0,5	0,5	2
AL-IV	B	15	30	3	0,5	0,5	0,5	2
AL-V	B	10	30	3	0,5	0,5	0,5	2

Table 4.2 Measurements of the Microhardness Values of the Composite Ni-Al-Cu

Composition name	$HV_{0,1}$
1st group (Sintered at 650 °C + Forged)	195 ± 12
1st group (Sintered only at 650 °C)	175 ± 10
2nd group (Sintered at 650 °C + Forged)	245 ± 15
3rd group (Sintered at 650 °C + Forged)	255 ± 10
4th group (Sintered at 650 °C + Forged)	260 ± 10
5th group (Sintered at 650 °C + Forged)	245 ± 10

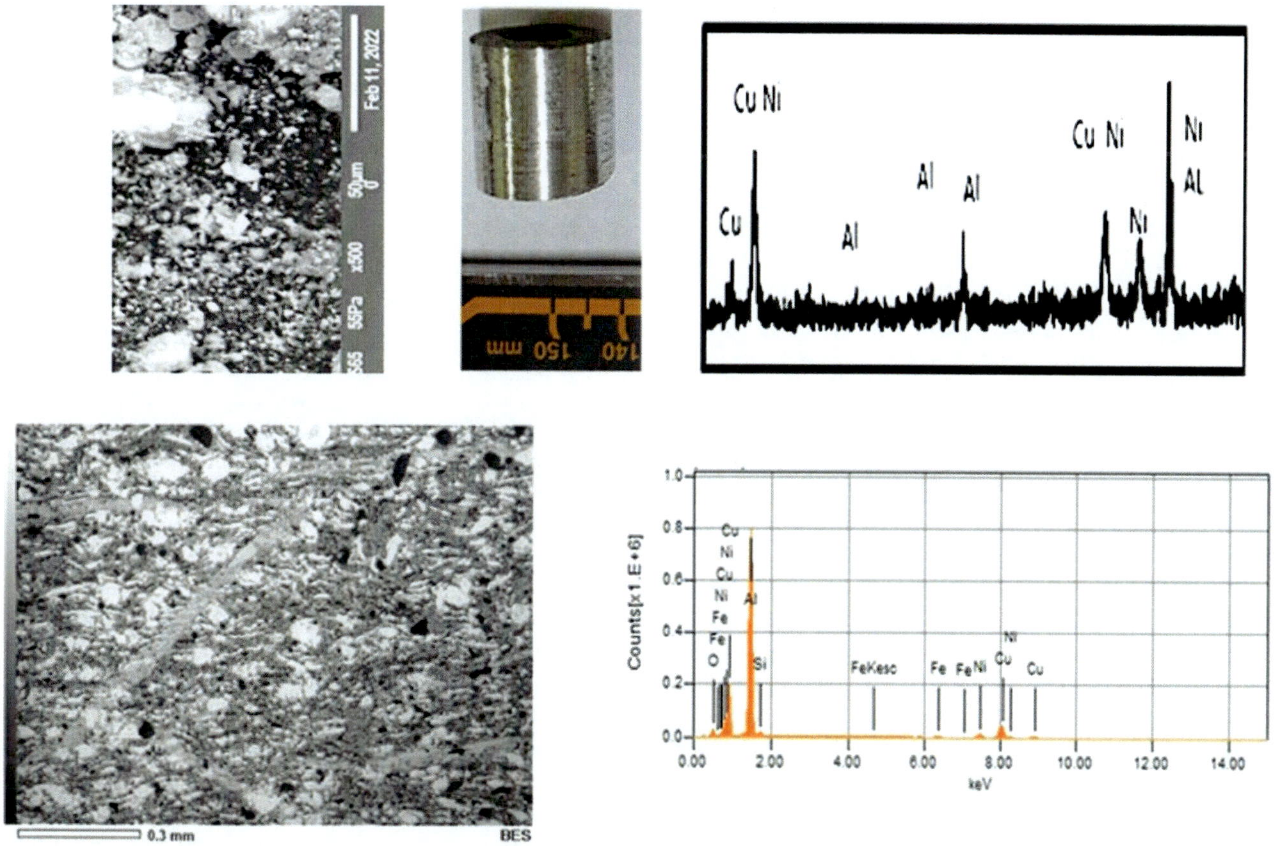

Fig. 4.1 Ni-Al-Cu intermetallic-based composite powder after 4 h milling and XRD and EDS chemical analyses with microstructure of the composites (AL-I) designed here from gas-atomized fresh scraps

Fig. 4.2 Sintered at 650 °C followed by hot forging of the specimens (*left*); compression test device on the Zwick test machine and damaged specimen 45°cracks (max. shear stress) after compression test

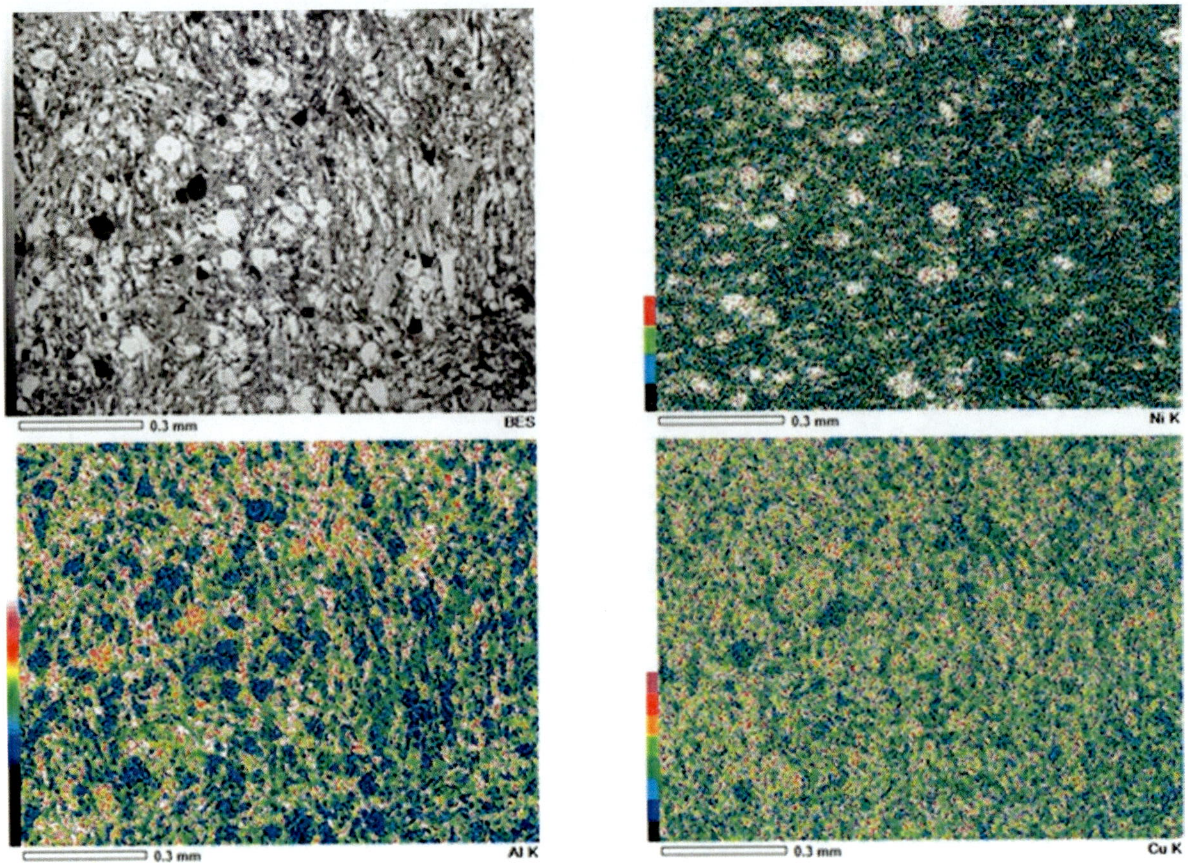

Fig. 4.3 Typical microstructural evaluation with elementary mapping analyses for only the composite AL-3

As for mechanical behaviour under the static and cyclic compression solicitations, very large experimental data have been achieved from these mechanical tests for each composition [17–20]. Figure 4.4 gives detailed information about these evaluations. It indicates static and cyclic compression test results obtained with the specimens sintered at 650 °C followed by hot forging. All the mechanical tests were carried out on the Zwick test machine. It seems that there is not so much difference between the static compression test results of these composites, lightly increase in ultimate tensile stress (UTS) levels in the AL-2, AL-3 and AL-4 because of the increase of Ni-Al in these compositions.

On the other hand, the cyclic-compression test (load–unload stress–strain curves) results gave a very low back stress appearance in the tested specimens. Very often, the load–unload stress evolution is not the same form, probably related to the Bauschinger effect that is seem a reasonable interpretation for this evolution. Although the initial stress levels of the materials have similar magnitude, evolution of the load stresses are different through the deformation and/or during the higher displacements.

Figure 4.4 gives fracture surfaces obtained from the damaged specimens solicited by cyclic compression tests. Damage analyses can be justified with these fracture surfaces. This means that the small amount of cumulative stresses occurred during the cyclic loading results in this type damage to the surfaces. Even this effect can be observed imperceptibly on the fracture surface of the other composites (Fig. 4.5).

Low-velocity impact (drop-weight) tests results are presented in Table 4.3. This type of test has been done by means of a drop tower to evaluate the response of the composites to dynamic loading. First, the force displacement curves were obtained for the compositions and then, the region under the load displacement curve was calculated for the absorbed energy for each composition. Energy absorbed on impact and energy absorption performance are presented in Table 4.3. These results were obtained under the laboratory scale and evidently the tests should be repeated to find an optimization in the results. Again, these values should be considered as indicative values.

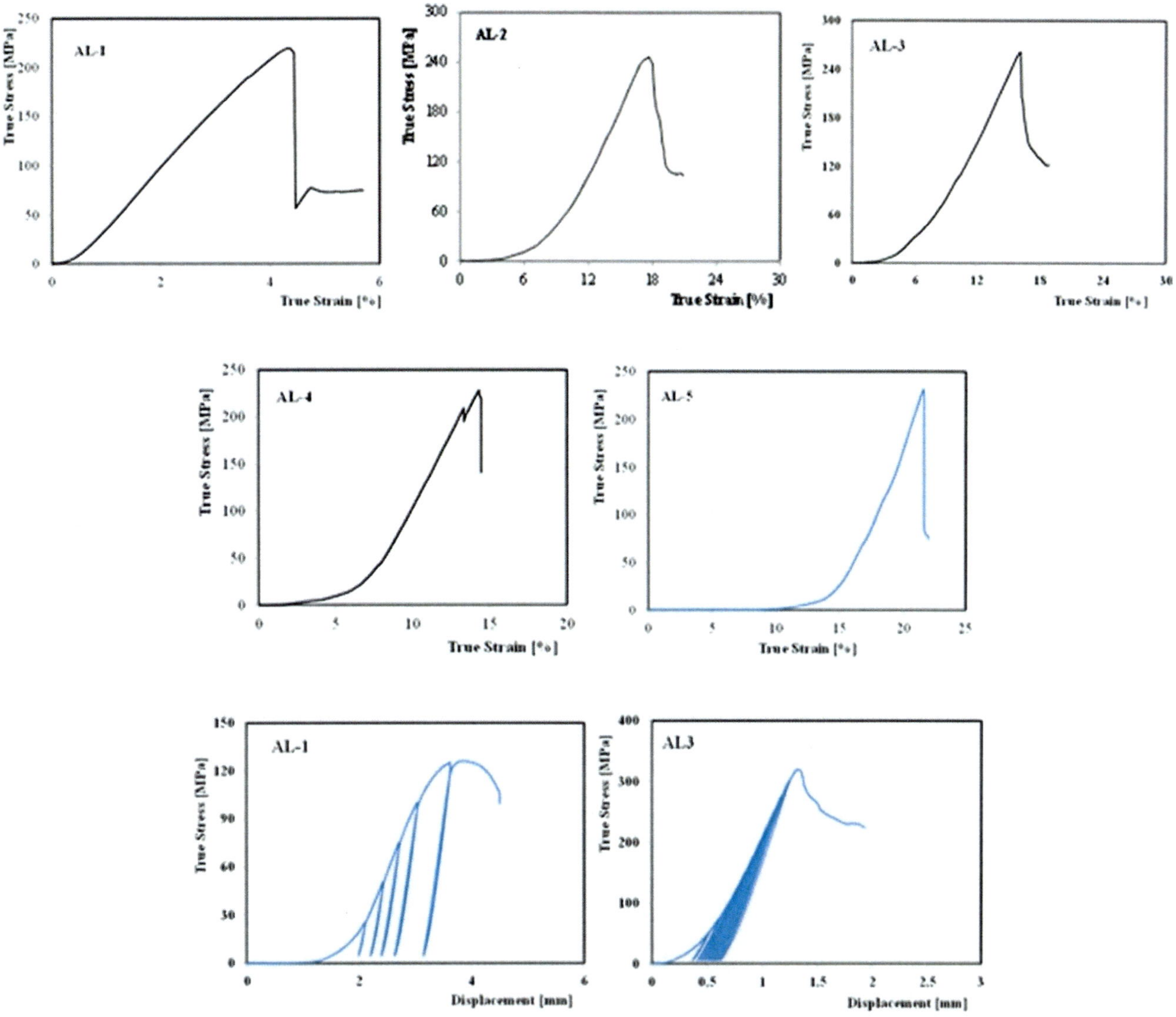

Fig. 4.4 Typical static- and cyclic-compression (load–unload curves) test results for the five compositions. All the tested samples were prepared in the same dimensions from the composites processed through the combined method: sintering + forging

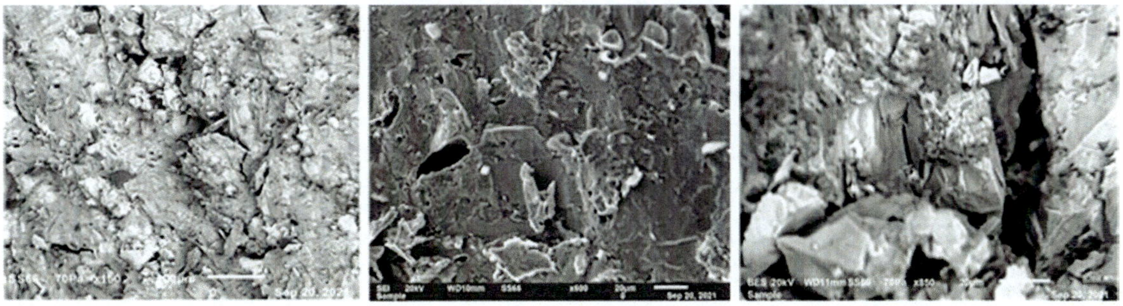

Fig. 4.5 Typical SEM fracture surfaces obtained from the damaged specimens solicited by cyclic compression tests

Table 4.3 Mean Values for the Energy Absorption Performance

Composite name	Impact energy, J	Absorbed energy, J	Energy absorption, %
AL-I	11,5	8.620	85.13
AL-2	11,5	8,985	86,12
AL-3	11,5	9.558	86.48
AL-4	11,5	10.168	89.50
AL-5	11,5	10.358	88.95

4.4 Conclusions

In the frame of a common research project, a new aeronautical composite was designed in five different compositions, called intermetallic-based composites, “Ni-Al-Cu”. These composites were intended to adjust easily for the new design of the parts used in aeronautical parts—for example, application of turbo compressor. Microstructural analysis has shown that a good bonding at the interface of matrix-reinforcement was essentially in the specimens manufactured with a combined process called sintering + forging—a tough and a healthy rigorous microstructure. Cyclic loading–unloading behaviours of these specimens produced with sintering + forging can give a certain behaviour similar to the Bauschinger effect that is seem as a main reason for this evolution.

References

1. Lu, Y.X., Tao, C.H., Yang, D.Z.: Improvement in mechanical properties of Nial matrix composites fabricated by reaction compocasting. Scr. Mater. **35**(10), 1243–1246 (1996). https://doi.org/10.1016/1359-6462(96)00276-x
2. Gu, D., Wang, Z., Shen, Y., Li, Q., Li, Y.: In-situ Tic particle reinforced Ti–Al matrix composites: powder preparation by mechanical alloying and selective laser melting behavior. Appl. Surf. Sci. **255**(22), 9230–9240 (2009). https://doi.org/10.1016/j.apsusc.2009.07.008
3. Popov, V.A., Shelekhov, E.V., Prosviryakov, A.S., Kotov, A.D., Khomutov, M.G.: Particulate metal matrix composites development on the basisof in situ synthesis of TiC reinforcing nanoparticles during mechanical alloying. J. Alloys Compd. **707**, 365–370 (2017)
4. Ferreira, L.F.P., Bayraktar, E., Robert, M.H., Miskioglu, I.: Chapter 17: Particles reinforced scrap aluminium based composites by combined processing sintering + thixoforming. In: Mechanics of Composite and Multi-functional Materials, Vol. 7, pp. 145–152. Springer, Berlin (2016) ISBN:978-3-319-41766-0
5. Lekatou, A., Karantzalis, A.E., Evangelou, A., Gousia, V., Kaptay, G., Gácsi, Z., et al.: Aluminum reinforced by WC and TiC nanoparticles (ex-situ) and aluminide particles (in-situ): Microstructure, wear and corrosion behaviour. Mater. Design. **65**, 1121–1135 (2015)
6. Bagheri, G.A.: The effect of reinforcement percentages on properties of copper matrix composites reinforced with TiC particles. J. Alloys Compd. **676**, 120–126 (2016)
7. Rudianto, H., Joo, J.K., Sun, Y.S., Jin, K.Y., Dlouhy, I.: Mechanical properties of sintered of Al-5.5Zn-2.5Mg-0.5Cu PM alloy. Mater. Sci. Forum. **794–796**, 501–507 (2014)
8. Gatamorta, F., Miskioglu, I., Bayraktar, E., Melo, M.L.: Recycling of aluminium-431 by high energy milling reinforced with Tic-Mo-Cu for new composites in connection applications. Mech. Composite Multi-Funct. Mater. **5**, 41–46 (2019). https://doi.org/10.1007/978-3-030-30028-9_6
9. Singh, S., Gupta, D., Jain, V., Sharma, A.K.: Microwave processing of materials and applications in manufacturing industries: A review. Mater. Manuf. Process. **30**, 1–29 (2015). https://doi.org/10.1080/10426914.2014.952028
10. Teer, D.G., Hampshire, J., Fox, V., Bellido-Gonzalez, V.: The tribological properties of MoS2/metal composite coatings deposited by closed field magnetron sputtering. Surf. Coat. Technol. **94**, 572–577 (1997)
11. Ferreira, L.M.P., Robert, M.H., Bayraktar, E., Zaimova, D.: New design of aluminium based composites through combined method of powder metallurgy and thixoforming. Adv. Mater. Res. **939**(1), 68–75 (2014)
12. Srivyasa, P.D., Charoo, M.S.: Role of fabrication route on the mechanical and tribological behavior of aluminium metal matrix composites—a review. Mater. Today Proc. **5**, 20054–20069 (2018)
13. Ferreira, L.F.P., Bayraktar, E., Robert, M.H., Miskioglu, I.: Chapter 17: Particles reinforced scrap aluminium based composites by combined processing sintering + thixoforming. In: Mechanics of Composite and Multi-functional Materials, pp. 145–152. Springer, Berlin (2016). ISBN: 978-3-319-41766-0
14. Mihlyuzova, L., Enginsoy, H.-M., Bayraktar, E., Slavov, S., Dontchev, D.: Tailored behaviour of scrap copper matrix composites reinforced with “Zn-Ni-Al”, low cost shape memory structures. In: Mechanics of Composite, Hybrid and Multifunctional Materials, Fracture, Fatigue, Failure and Damage Evolution, Vol. 3, pp. 49–54 (2021)
15. Mihlyuzova, L., Enginsoy, H.M., Dontchev, D., Bayraktar, E.: Tailored behaviour of scrap copper matrix composites reinforced with zinc and aluminium: low cost shape memory structures. Mech. Composite Hybrid Multifunct. Mater. **6**, 27–34 (2020)
16. Kai, X.Z., Tian, K.L., Wang, C.M., Jiao, L., Chen, G., Zhao, Y.T.: Effects of ultrasonic vibration on the microstructure and tensile properties of the nano ZrB2/2024Al composites synthesized by direct melt reaction. J. Alloy Comp. **668**, 121–127 (2016)
17. Gatamorta, F., Enginsoy, H.M., Bayraktar, E., Miskioglu, I., Katundi, D.: Design of recycled alumix-123 based composites reinforced with γ-Al_2O_3 through combined method; Sinter + forging. Mech. Composite Hybrid Multifunct. Mater. **6**, 9–18 (2020)

18. Koli, D.K., Agnihotri, G., Purohit, R.: Advanced aluminium matrix composites: the critical need of automotive and aerospace engineering fields. Mater Today Proc. **2**(4–5), 3032–3041 (2015)
19. Bayraktar, E., Gatamorta, F., Enginsoy, H.M., Polis, J.E., Miskioglu, I.: New design of composites from fresh scraps of niobium for tribological applications. Mech. Composite Hybrid Multifunct. Mater. **6**, 35–44 (2020)
20. Wang, P., Gammer, C., Brenne, F., Niendorf, T., Eckert, J., Scudino, S.: A heat treatable TiB2/Al-3.5Cu-1.5Mg-1Si composite fabricated by selective laser melting: microstructure, heat treatment and mechanical properties. Composites Part B. **147**, 162–168 (2018)

Check for updates

Chapter 5
Multifunctional Behaviour of Titanium-Based Composites Reinforced with Recycled Iron and Magnetic Iron Oxide (Fe_3O_4): Magnetic Shape Memory Composites

Stanislav Slavov, Ivelina Ivanova, Virginia Alexandrova, Dimitar Dontchev, and Emin Bayraktar

Abstract Titanium-based composites have potential applications in the aeronautical industry as multifunctional materials and/or magnetic shape memory composites (MSMCs). In this work, gas-atomized recycled pure iron (1–5 μ) and fine magnetic iron oxide (<1 μ-Fe_3O_4) were used as reinforcements for pure titanium (VWR) Two types of composites were fabricated by using a powder metallurgy-combined method: sintering + forging. Two types of processes were carried out at two different sintering temperatures (600 °C and 850 °C). As known, the Ti-Fe eutectic temperature is ~750 °C. Increasing of the pure iron to 45 wt% gives rise to the formation of Ti-Fe intermetallic in the composites. As a result, mechanical properties of the sintered forging composites were considerably improved. Electrical and thermal behaviour and magnetic permeability were measured. The microstructure analyses and the diffusion mechanism at the Ti-Fe interface are also investigated by using scanning electron microscopy (SEM).

Keywords Titanium · Sintering + forging · Recycled iron · Magnetic Fe_3O_4 · Shape memory

5.1 Introduction

This work is related to the "magnetic shape memory composites" (MSMCs) that are an exceptional titanium–iron (Ti-Fe) intermetallic alloy-based composite reinforced with magnetic iron oxide (Fe_3O_4). They have multifunctional properties. They show an adjustable solid-to-solid phase transformation through thermal and/or mechanical loadings that may result in large recoverable, rigid strains of about 1–8% [1, 2]. The formulation of Ti-45Fe–10Fe_3O_4 is a face centre cubic structure (i.e., high symmetry, FCC) at high temperature and low stress levels; however, they have a martensitic structure and are exceptionally accommodating though retaining the strength of metals and can convert thermal energy into mechanical work.

For this reason, design of the MSMCs in the form of a composite structure is a very useful process. Today, there is an increasing demand in aeronautical, aerospace (e.g., drones), automotive, biomedical, energy conversion and storage-related applications of MSMCs composites [3–17]. These composites are basically ferromagnetic shaped memory alloys (FSMAs); they are future energetic multifunctional materials for possible applications of high-power–lightweight actuators controlled by magnetic fields [9–17].

A new family of magnetic memory shape–smart materials can be created with extraordinary strains up to 15–20% and frequencies in the range of 1–2 kHz. The magnetic shape memory actuator is a potential device of MSMCs, which can achieve high-performance electromagnetic actuation by using the properties of MSM alloys (1–5, 8–15–26).

The present work described in this chapter is an ongoing common research project, as a new alternative design for "low-cost composites" from ferromagnetic shape memory alloy composite (MSMAC) that have been developed.

S. Slavov · I. Ivanova · V. Alexandrova · D. Dontchev
University of Chemical Technology and Metallurgy (UTCM), Sofia, Bulgaria
e-mail: stanislavslavov@uctm.edu; dontchev@uctm.edu

E. Bayraktar (✉)
School of Mechanical and Manufacturing Engineering, ISAE-SUPMECA-Paris, Paris, France
e-mail: emin.bayraktar@isae-supmeca.fr

V. Chalivendra, F. Gardea (eds.), *Mechanics of Composite, Hybrid & Multi-functional Materials, Volume 5*, Conference Proceedings of the Society for Experimental Mechanics Series, https://doi.org/10.1007/978-3-031-17445-2_5

5.2 Experimental Conditions

In this work, gas-atomized fresh scrap recycled pure iron (1–5 μ) and fine magnetic iron oxide (<1 μ-Fe_3O_4) were used as a reinforcement for pure titanium (VWR) Two types of composites were processed by using a powder metallurgy-combined method: sintering + forging. Two types of processes were carried out at two different sintering temperatures (600 °C and 850 °C). As known, Ti-Fe eutectic temperature is ~750 °C. Figures 5.1, 5.2 and 5.3 (see Sect. 5.2.1) indicate clearly the mutual diffusion and the evolution of Ti-Fe in a binary phase diagram.

A homogeneous mixture of the powder was carried out by high-energy milling in a planetary ball milling with an inert argon atmosphere to prevent oxidation of the powders. A little bit of pure nano-electrolytic copper (2 wt%) powder was added to homogenize the matrix during the milling and increase the chemical bonding diffusion between the matrix and the reinforcements. Sintering was done under argon inert atmosphere according to the formulation of an aeronautic company.

Final forging was carried out on the sintered specimens at the temperature lower than that of sintering. Increase of the pure iron addition to 45 wt% gives progress to the formation of Ti-Fe intermetallic in the composites. As a result, mechanical properties of the sintered forging composites were considerably improved. Electrical conductivity and the capacity of the magnetic permeability were measured.

Measurements of magnetic properties, basically magnetization and hysteresis loop of the specimens were performed at room temperature using a special type of commercial GMW-vibrating sample magnetometer, VSM 3474-140. This supplies the coercive force, remanence, saturation magnetization, and so forth, as well as gives information about the magnetization process.

Electrical measurements were performed on the 3–4 cylindrical samples with a diameter of 21 mm and a thickness of about 7 mm. The surfaces of the samples were plane–parallel polished carefully, followed by coating with Ag paste and then dried. The measurements of the conductivity as a function of frequency (100 Hz–100 kHz) were carried out by using a "ZANNER" CM6 impedance analyzer with an error rate of ≤2%. The resistances, R, were determined at various frequencies, assuming an equivalent circuit consisting of RC in parallel. All the measurements were made at the physical laboratory of UTCM-Sofia, Bulgaria.

The results were given under laboratory conditions used in the present work. Naturally, the values have been given as a mean value for the specimens containing the same composition. Evidently, microstructure and the form, microporosity, and the surface conditions of each sample can easily influence the measurements. That is the reason that all the specimens were attempted to be produced under the same conditions. These results should be considered as indicative, and it seems that form and surface conditions should be improved again.

The microstructure analyses have been carried out by SEM with a mapping and "EDS" chemical analyses. The density measurements of the four specimens were carried out by using the Archimedes method. The values change between 4.75 and 6.18 g/cm³. All the mechanical tests have been carried out on a Zwick mechanical test device installed at the laboratory of ISAE-SUPMECA-Paris, France.

Three cylindrical specimens (H/D ≥ 1.5) were tested under quasistatic- and cyclic-compression conditions according to the DIN 50106 norm by using a constant test speed as 1,5 mm/min. Finally, three-point bending, 3PB, tests were carried out on the same mechanical test system. Low-velocity impact (drop-weight) tests were carried out with a drop tower to evaluate

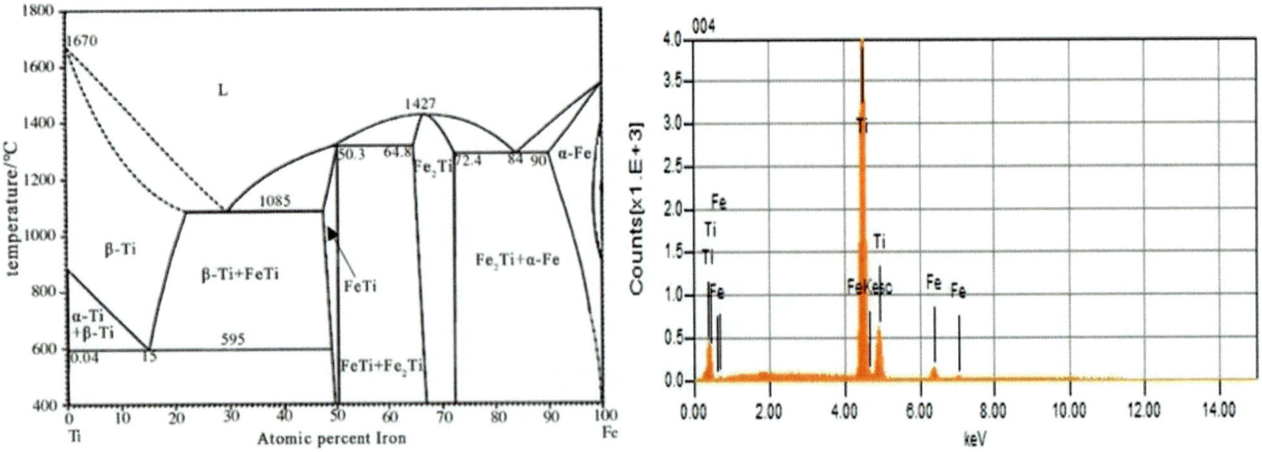

Fig. 5.1 Ti-Fe binary phase diagram and microstructure of the Ti-45Fe–10Fe_3O_4 and EDS shown in this group

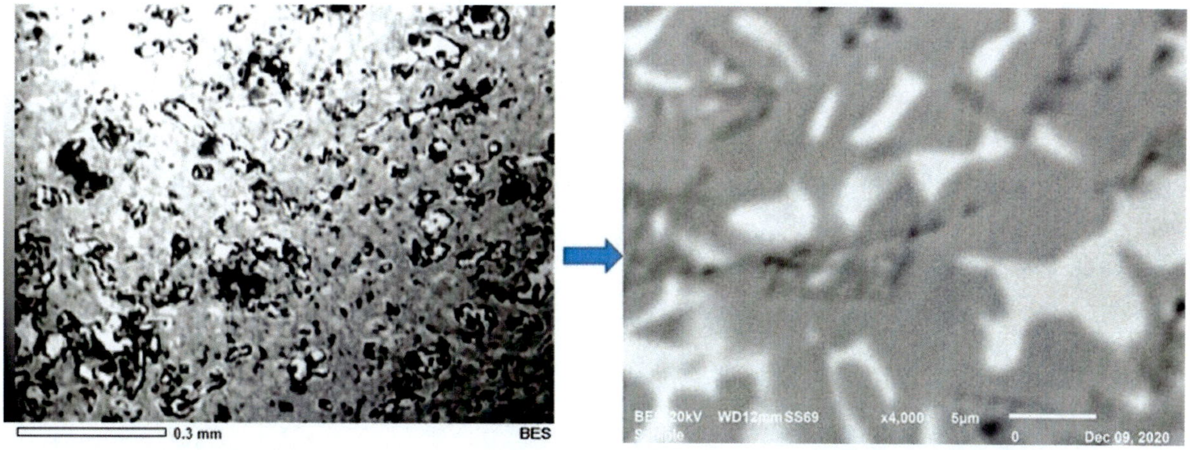

Fig. 5.2 General microstructure of the Ti-45Fe–10Fe_3O_4 composite and detail of the structure, distribution of Fe and Fe_3O_4 in the titanium matrix in this group

the response of the composites to dynamic loading. Three cylindrical specimens (H/D $\geq$ 1.5) were also used under low-velocity/drop-weight test conditions. Microhardness ($HV_{0.1}$) tests were conducted on the polished and etched specimens. The mean microhardness values are presented in Table 5.1 with ±15–20% accuracy for the two manufacturing processing techniques, respectively.

5.2.1 *Microstructure and Mapping Analyses of the Compositions Ti-45Fe–10Fe_3O_4 Produced by the Sintering + Forging Process*

General microstructure of the composite, Ti-45Fe–10Fe_3O_4, "MSMC" is presented in Figs. 5.1, 5.2 and 5.3. "EDS" chemical and mapping analyses have also been evaluated together by using the Ti-Fe binary phase diagram. Distribution of reinforcement elements is generally homogenous. A small amount of agglomeration of the secondary minor reinforcements was observed with Fe_3O_4, nano-magnetic iron oxide (10 wt%) in the structure even if a doping was done before the mixture of the matrix.

These elements are localized near the grain boundaries as small-size clusters. This case is directly related of mixture processes. All the specimens were prepared under laboratory conditions; therefore, the results should show improved operational conditions. Distribution of the major reinforcement elements is homogenous. This case should be related to the sintering + forging process.

Figure 5.4 shows dynamic scanning calorimetry (DSC) results for the composite developed in this work (i.e., Ti-45Fe–10Fe_3O_4 composite) that show a transformation point at the level of 252–255 °C depending on the time and heating process. As is known, DSC is used for a thermal analysis technique in which the heat flow into or out of a sample is measured as a function of temperature or time, while the sample is exposed to a controlled temperature program as in the standard and in the users' manual as indicated. Heating temperature was kept up to 300 °C with a heating rate of 10 °C/min Heating areas were measured as 95 mJ and ΔH = 3,5 J/g. The measurement was carried out on the 50 mg of the sample taken at the center of the specimen.

5.2.2 *Measurements of Electrical Properties*

Electrical measurements were performed on the 3–4 cylindrical samples with a diameter of 21 mm and a thickness of about 7 mm. The surfaces of the samples were plane–parallel polished carefully, followed by coating with Ag paste and dried. The measurements of the conductivity as a function of frequency (100 Hz–100 kHz) were carried out by using a "ZANNER" CM6 impedance analyzer with an error rate of $\leq$ 2%. The resistances, R, were determined at various frequencies, assuming an equivalent circuit consisting of RC in parallel.

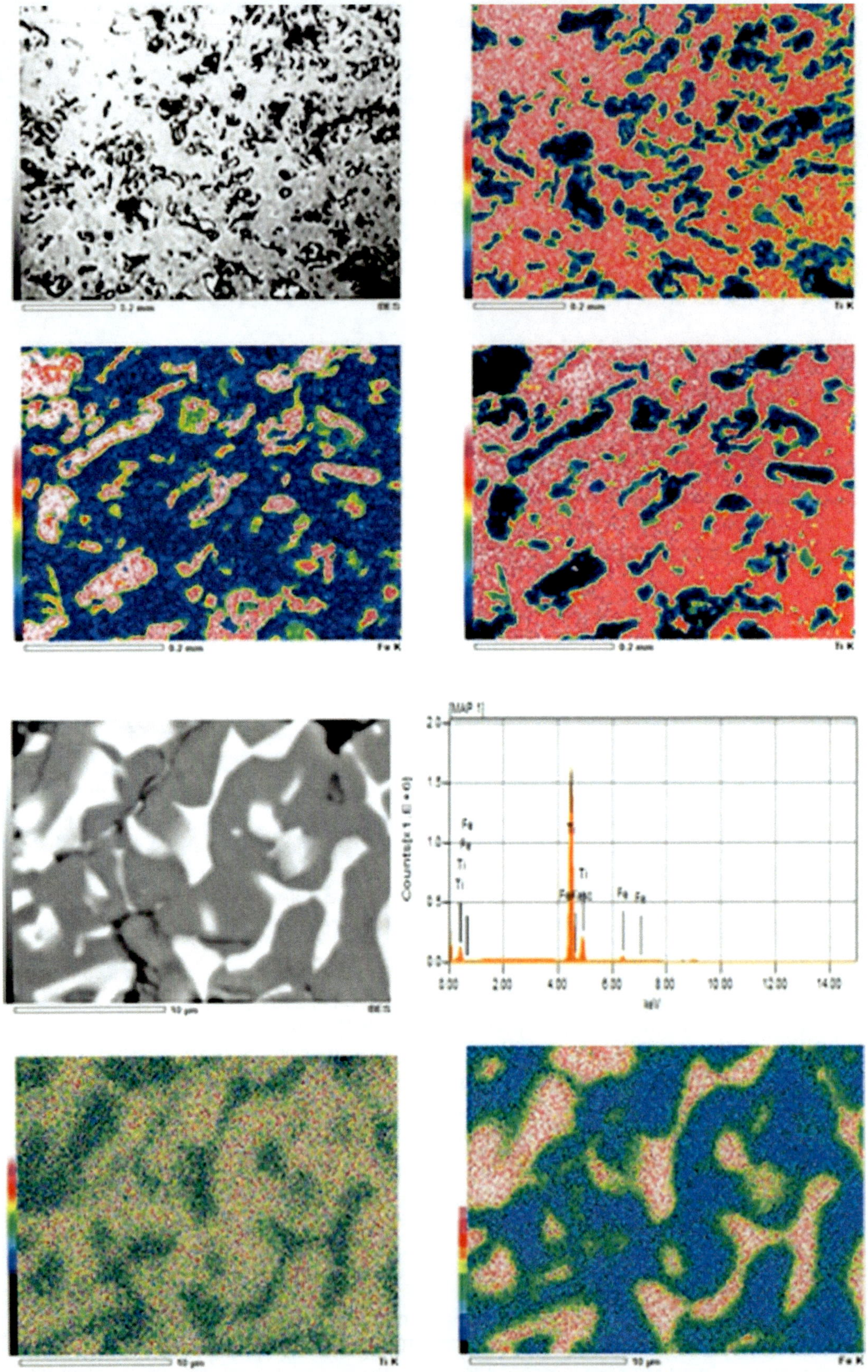

Fig. 5.3 General microstructure of the Ti-45Fe–10Fe_3O_4 composite, with mapping analyses and evaluation of the distribution of each element reinforcements

Table 5.1 Measurements of the Microhardness Values of the Composite Ti45Fe–Fe_3O_4

Composition name	Micro hardness values, $HV_{0,1}$ (at interface only)
Ti45Fe–Fe_3O_4 (Sintered at 600 °C)	330 ± 15
Ti45Fe–Fe_3O_4 (Sintered at 850 °C)	405 ± 20

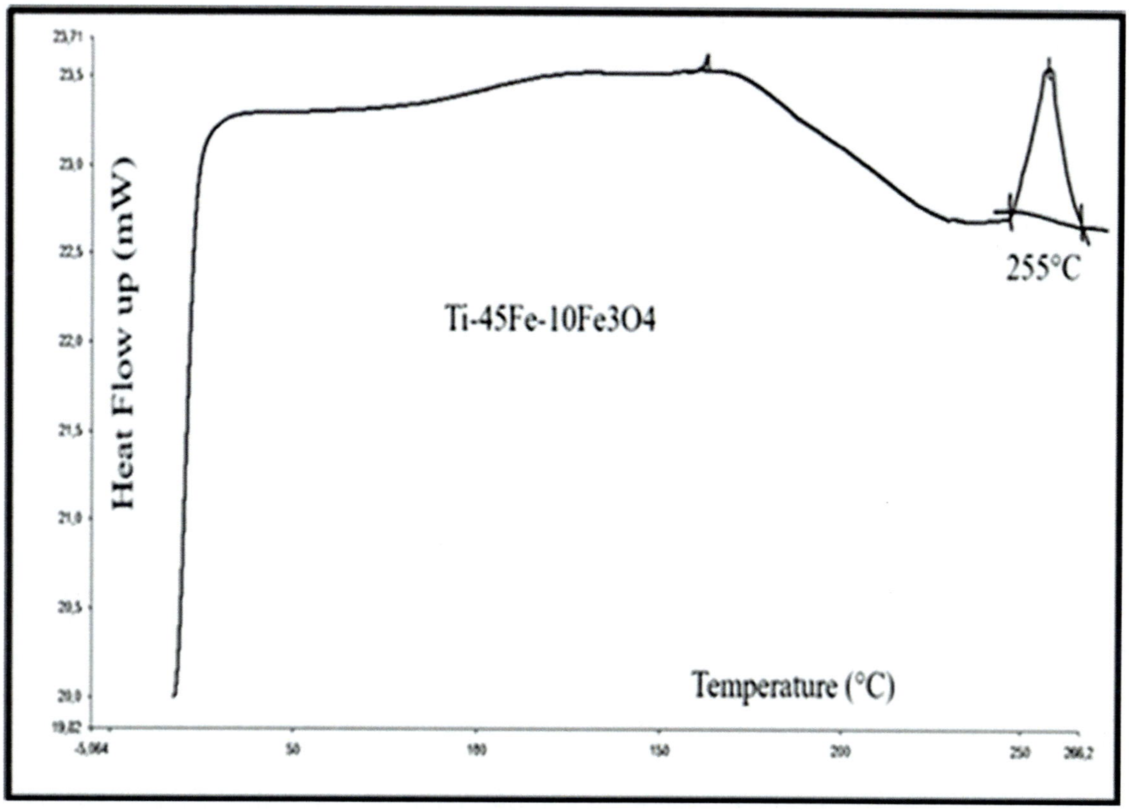

Fig. 5.4 DSC diagram measured for the Ti-45Fe–10Fe_3O_4 composite heated up to 300 °C, a transformation point at the level of 252–255 °C

All the measurements were made at the physical laboratory of UTCM-Sofia, Bulgaria. Figure 5.5 gives all the mean electrical conductivity measurements as a function of the frequency taken from 3–4 thin specimens prepared according to the standard indicated in the user manual for the device.

The probable structure of the Ti-Fe intermetallic alloy is composed of Ti ions surrounded by a dense shell of Fe atoms, forming a cubic crystallographic structure. These individual cubic phase structures are grouped into inhomogeneously distributed microforms among a Ti-Fe nanostructure of the matrix. Depending on the time of thermal treatment of the alloy, as well as the pressure, these formations vary in size and shape. Because of the presence of Ti-Fe phases with various priority crystallographic orientations, the role of the Ti atom in the electron–diffusion processes [XX1] will be decisive for the interpretation of the electrical properties. According to the same authors, the structure of the electronic levels of Ti-Fe is determined depending on the lattice constant, which varies from 2.8–3A°.

In this context, the difference in energy-filling levels (i.e., Fermi level) for titanium and iron atoms (higher for titanium and lower for iron, respectively) commented by other authors [XX2–4] can be related to the conductivity phenomena of the sample. As the frequency of the applied external electric field increases, the population of electrons in the electron levels of the two atoms changes to varying degrees. Given that the cubic structure of the Ti-Fe phases are the main elements for conductivity, the various field energies lead to the occurrence of metal-to-metal contact phenomena.

Consequently, along a certain crystallographic axis of the Ti-Fe phases, thanks to the electron–diffusion ability of the titanium atom and the state-excited states of the atoms, zones with a priority direction of conductivity are formed. These zones, formed as nanocubic structures with different orientations, form polarization clusters under the action of the alternating electric field. For miles and micro-sized volumetric parts of the sample, grouping these clusters will result in charge compensation and thus reduced conductivity.

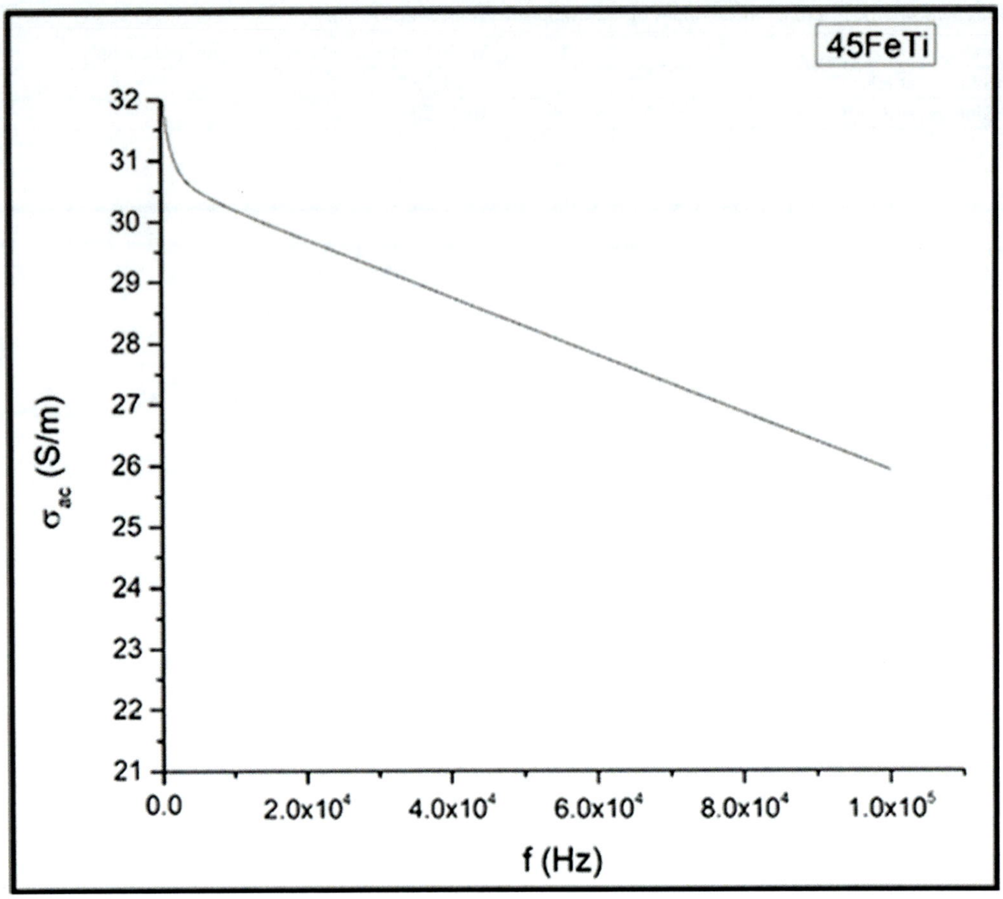

Fig. 5.5 Measurement of the electrical conductivity (S/m) as a function of the frequency for the composite, Ti45Fe–10Fe_3O_4

5.2.3 Time-Dependent Behaviour by Means of Nanoindentation: Wear-and-Creep Tests

For these tests, the creep compliance and the stress exponent were calculated by using data collection, as defined in Eq. 5.1 [13]:

$$\varepsilon(t) = \sigma_0 J(t) \tag{5.1}$$

where σ_0 is the constant stress applied and $J(t)$ is calculated using Eq. 5.2:

$$J(t) = A(t)/(1-v)P_0 \tan\theta \tag{5.2}$$

In Eq. 5.2, $A(t)$ is the contact area; P_0 is the constant applied load; θ is the effective cone angle, which is 70.3° for a Berkovich indenter; and the Poisson's ratio ν is assumed to be 0.3. This approach considers how the contact area under the Berkovich tip alters while displacement into the surface changes.

The strain-versus-time behaviour during creep is characterized by a high-strain rate in the primary stage of creep and then in the secondary, steady-state stage of creep; the strain rate is given in Eq. 5.3 and written as:

$$\dot{\varepsilon} = K\sigma^n \tag{5.3}$$

where K is a constant and n is the stress exponent. The strain rate is calculated in the software and in turn n is obtained from the log–log plot of strain rate versus stress in the secondary stage of creep. The materials under consideration are heterogeneous in nature and the fact that the nanoindentation test is carried out over a small area/volume, a large scatter in the data is observed, and to overcome this, a sampling number was taken as large as possible.

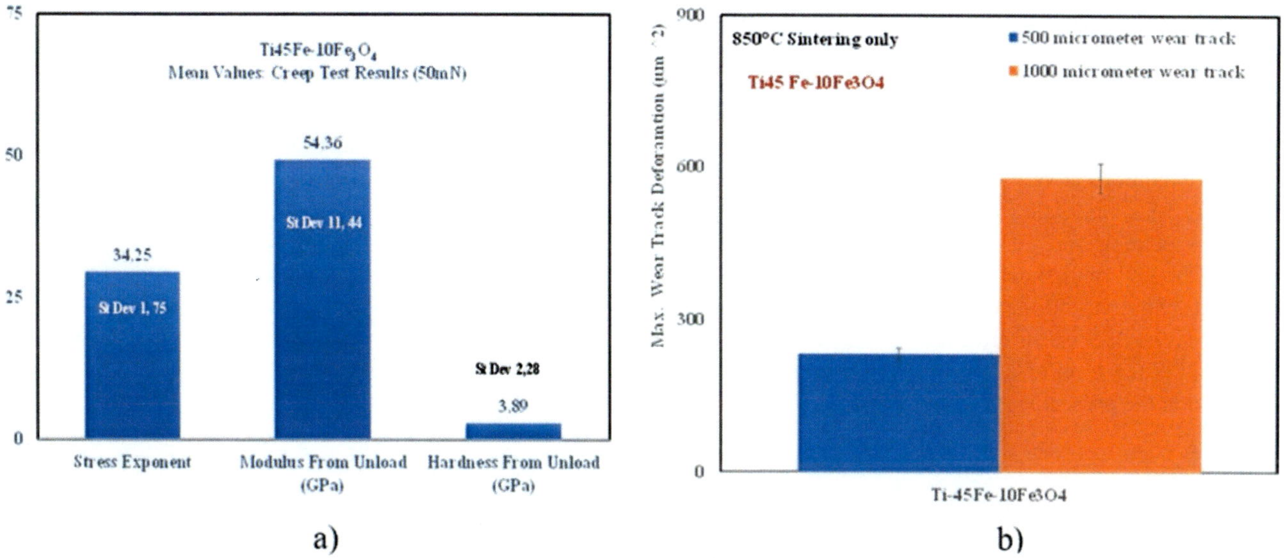

Fig. 5.6 (**a**) Mean creep test results carried out under the load of 50 mN for the Ti45Fe–Fe_3O_4 specimens and (**b**) maximum wear track deformation measured as the area between the initial profile and the final (residual) profile of the wear tracks for 500 and 1000 μm long-wear paths for the Ti-45Fe–Fe_3O_4 specimens

Figure 5.6 shows mean creep test results carried out under the load of 50 mN for the Ti45Fe–Fe_3O_4 specimens. Stress exponent, modulus from unload (GPa) and hardness from GPa values, were shown as the mean values obtained for 10-creep experimental results. Figure 5.6b shows the maximum wear track deformation measured as the area between the initial profile and the final (residual) profile of the wear tracks for 500 and 1000 μm long-wear paths for the Ti-45Fe–Fe_3O_4 specimens. These are the averages of the 10 wear tests for each type of manufacturing composites. Finally, all the observations indicate that the prepared samples are more resistant to wear.

5.2.4 Mechanical Testing Under Static- and Dynamic-Loading Conditions

Certain mechanical tests have been carried out for the composite Ti45Fe–Fe_3O_4. All the test results were evaluated under the laboratory levels for the mechanical behaviour of this composite produced with the combined method: sintering + forging processes. Obviously, this idea should be developed thoroughly on this composite for the piece in the industrial scale. In the figures, all the values were taken as the mean values tested at least four of the same type of samples. It seems that the composition designed and proposed for a special application of the electronic parts for an alternative replacement for actuators generally show a ductile behaviour.

Quasistatic 3-Point Bending, 3PB Testing

Figure 5.7 gives the static 3Point Bending, 3PB, test results as a mean value obtained on two different samples obtained with two different sintering temperatures, 600 and 850 °C specimens, respectively. All the tests have been carried out at ASTM790 with the distance of 32 mm between two points.

It seems that two different processes (here only sintering temperature 600 and 850 °C for the same compositions) influence the toughness of the composite very convincingly. Many experimental results gave the same results here for at least four samples for each sintering process. The composition designed and proposed for a special application of the electronic parts for an alternative replacement for actuators generally show a ductile behaviour.

Cyclic-Compression Testing

Many samples were prepared for the same composite, Ti45Fe–Fe_3O_4, MSMC for the cyclic-compression tests. In fact, this type of composite that the authors call a magnetic shape memory composite (MSMC) absolutely it should be known, whether there is a back stress after each deformation cycle essentially at the lower stress levels. In fact, these parts are subject to the cyclic solicitations at very low frequencies under the service conditions.

Figure 5.8 gives loading and unloading displacements of the specimens under cyclic compression for the Ti45Fe–10Fe_3O_4, MSMC specimen sintered at 600 °C (*left*) and 850 °C (*right*), respectively. These tests have been carried out in two samples groups; the first ones sintered at 600 °C and the second ones sintered at 850 °C. For both of the test groups, the first stress level was chosen as 50 MPa and at each step, a stress increment was chosen as 25 MPa.

There is no back-stress effect or other residual stresses on the samples. Both types of the samples have given a transformation point at the level of 2 mm of the displacements for the magnetic-shape memory composites—Ti45Fe–Fe_3O_4, MSMC designed in the present work.

5.2.5 *Low-Velocity Impact Drop-Weight Testing*

Low-velocity or dynamic compression tests results are given in Fig. 5.9 for the Ti45Fe–10Fe_3O_4, MSMC specimen. Maximum force was evaluated there by the values from both support data points. Here, a series of impact tests were performed at room temperature using the instrumented drop-weight test device, as explained in detail in Sect. 5.2.

Three specimens were tested for this composition. As known, an impact resistance is related to absorbed energy. All the specimens have shown that a major part of the impact force is used to keep the balance with the inertia force, and only a small portion of the impact force is used to damage and fracture the specimen. The microstructure of the tested specimens has shown a dynamic behaviour of this composite and gives hopeful data as a ductile behaviour in case of impact with a low velocity.

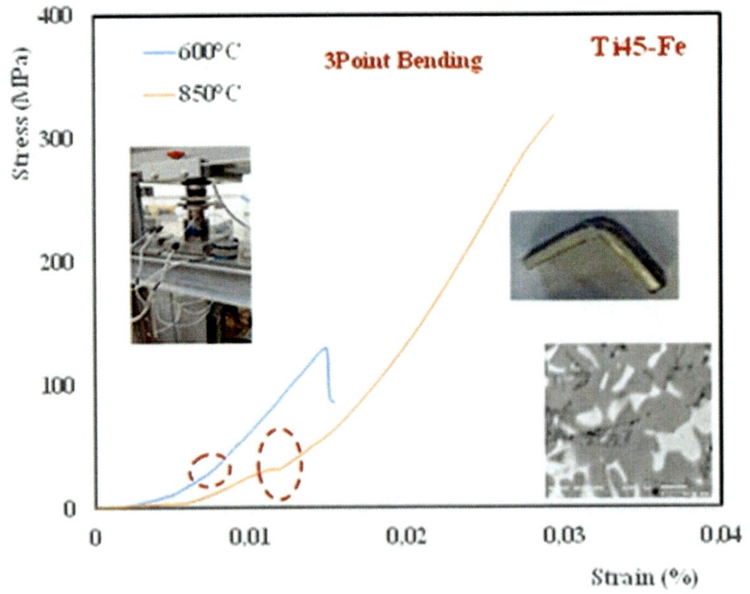

Fig. 5.7 Static 3PB tests (as a mean value) for the Ti45Fe–Fe_3O_4, MSMC specimen

Table 5.2 Basic information taken from magnetic permeability test of the Ti-45Fe–Fe_3O_4–MSMC specimens (as a mean value measured separately on 3–4 specimens)

Ms (emu/g)	Mr (emu/g)	Hc (G)	χ_m	Hr (G)
46, 73	0, 87	160, 39	1, 10E-03	2920

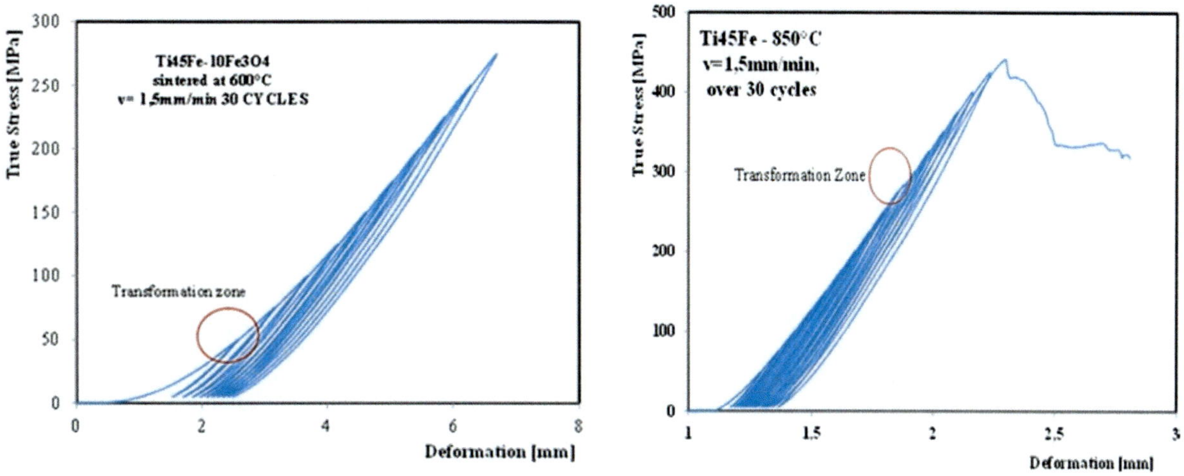

Fig. 5.8 Loading and unloading displacements of the specimens under cyclic compression for the Ti45Fe–10Fe_3O_4, MSMC specimen sintered at 600 °C (*left*) and 850 °C (*right*), respectively

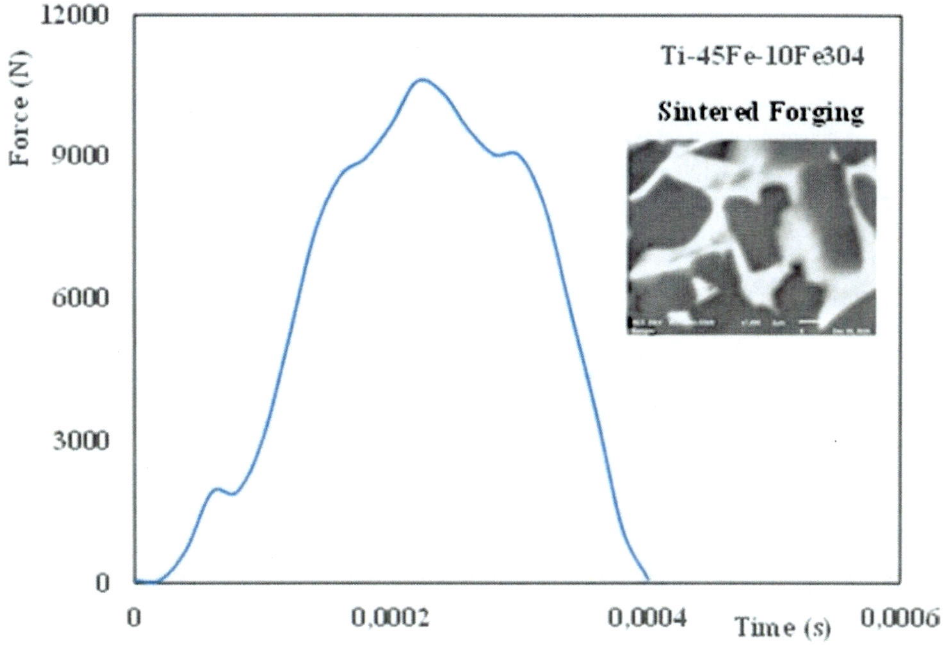

Fig. 5.9 Low-velocity impact behaviour for a Ti45Fe–10Fe_3O_4, MSMCs composite obtained by dynamic compression test

5.2.6 Magnetic Permeability Measurements of the Ti45Fe–Fe_3O_4–MSMCs Materials

Magnetic permeability measurements of the Ti45Fe–Fe_3O_4–MSMCs have been carried out under laboratory conditions. Basically, magnetization and a hysteresis loop of the specimens were completed at room temperature using a special type of commercial GMW vibrating sample magnetometer, VSM 3474-140. This supplies the coercive force, remanence, saturation magnetization, and so on and gives information about the magnetization process (Fig. 5.10).

A high-level soft-magnetic permeability was obtained on this composite. These magnetic permeability values give potential applications in the aeronautical area for this composite. The parameters Mrs, Ms, Hc and Hcr are dependent on a grain-size ferromagnetic and can provide information about magnetic domain structures. The magnetic susceptibility (designated by χ_m) measures the ability of a material to magnetize under the action of a magnetic stimulation—of a magnetizing field—to which it is submitted (Table 5.2). The values have been calculated from the diagrams shown earlier in Figs. 5.8 and 5.9.

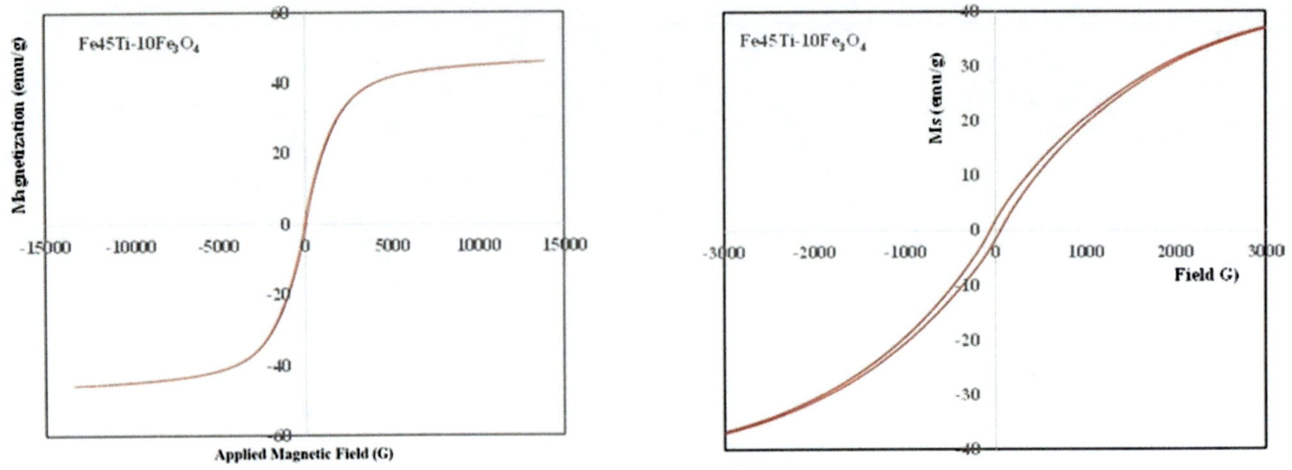

Fig. 5.10 Magnetic permeability graph with a detail of hysteresis of the Ti45Fe–Fe_3O_4–MSMC specimen

5.3 Conclusions

A new design of magnetic shape memory composite, Ti-45Fe–10Fe_3O_4, MSMCs was presented. This composite can be used easily and adapted as a spherical actuator for spacecraft and/or for aircraft engine attitude controls; and as presented here, relies on a simple dipole magnet to exchange momentum with the spacecraft. Microstructural analysis has shown that a good bonding at interface of matrix-reinforcement is essential in the specimens manufactured with a combined sintering + forging process—a tough and sound microstructure. Incremental cyclic-loading behaviours of these specimens produced with sintering + forging can give a certain behaviour to shape memory alloys. Electrical conductivity and magnetic permeability give very hopeful values for future work in the production in aeronautical applications.

Acknowledgements The authors want to thank financial support from French Consulate in Sofia, Bulgaria–UTCM, University of Chemical Technology and Metallurgy. All the mechanical tests were carried out on a Zwick mechanical test device installed at the laboratory of ISAE-SUPMECA-Paris, France, and we acknowledge the company for their technical help. We acknowledge the French Aeronautical Society, Airbus-Helicopters, Dr. G. Zambelis for his support and technical help for this project. We acknowledge Prof. I. Miskioglu from Michigan Technical University, MI-USA, for the nanoindentation tests for the creep and wear behaviour of the composite studied in this project.

References

1. Canadinç, D., Trehern, W., Ma, J., Karaman, I., Sun, F., Chaudhry, Z.: Ultra-high temperature multi-component shape memory alloys. Scr. Mater. **158**, 83–87 (2019)
2. Ferreira, L.-M., Bayraktar, E., Miskioglu, I., Robert, M.-H.: Design of magnetic aluminium (AA356) composites (AMCs) reinforced with nano Fe_3O_4, and recycled nickel: copper particles. Mech. Compos. Hybrid Multifunct. Mater. Springer. **5**, 93–100 (2018)
3. Yazdani, H., Hatami, K., Khosravi, E., Harper, K., Grady, B.P.: Strain-sensitive conductivity of carbon black-filled PVC composites subjected to cyclic loading. Carbon. **79**, 393–405 (2014)
4. Khalil, W., Saint-Sulpice, L., Arbab Chirani, S., Bouby, C., Mikolajczak, A., Ben Zineb, T.: Experimental analysis of Fe-based shape memory alloy behaviour under thermomechanical cyclic loading. Mech. Mater. **63**, 1–11 (2013)
5. Nespoli, A., Besseghini, S.: A complete thermo-mechanical study of a NiTiCu shape memory alloy wire. J. Therm. Anal. Calorim. **103**, 821–826 (2011)
6. Balak, Z., Abbasi, S.M.: Influence of the Ti content, training cycles and pre-strain on the two-way shape memory effect in NiTi alloys. Mater. Des. **32**, 3992–3996 (2011)
7. Bayraktar, E., Ayari, F., Tan, M.J., Bayraktar, A.T., Katundi, D.: Manufacturing of aluminium matrix composites reinforced with iron oxide (Fe_3O_4) nanoparticles: microstructural and mechanical properties. Metallurg. Mater. Trans. B Springer. **45**(2), 352–362 (2014)
8. Benefan, O., Brown, J., Calkins, F.T., Kumar, P., Stebner, A.P., Turner, T.L., Vaidyanathan, R., Webster, J., Young, M.L.: Shape memory alloy actuator design: CASMART collaborative best practices and case studies. Int. J. Mech. Mater. Design. **10**, 1–42 (2014)
9. Hartl, D.J., Lagoudas, D.: Aerospace applications of shape memory alloys. Proc. Inst. Mech. Eng. G J Aerosp. Eng. **221**(Special Issue), 535–552 (2007)
10. Bellouard, Y.: Shape memory alloys for microsystems: a review from a material perspective. Mater. Sci. Eng. A. **481-482**, 582–589 (2008)
11. Ferreira, L.-M., Miskioglu, I., Emin Bayraktar, D., Katundi.: Reinforcement effect of nano Fe_3O_4 and Nb2Al on the mechanical and physical properties of Cu-Al based composites. Mech. Compos. Hybrid Multifunct. Mater. Springer. **5**, 101–108 (2018)

12. Ferreira, L.-M.-P., Bayraktar, E., Miskioglu, I., Robert, M.-H.: New magnetic aluminum matrix composites (Al-Zn-Si) reinforced with nano magnetic Fe_3O_4 for aeronautical applications. Adv. Mater. Process Technol. **4**(3), 358–369 (2018)
13. Ferreira, L.-M.-P., Bayraktar, E., Miskioglu, I., Robert, M.-H.: Influence of nanoparticulate and fiber reinforcements on the wear response of multiferroic composites processed by powder metallurgy. Adv. Mater. Process Technol. **3**(1), 23–32 (2016)
14. Ferreira, M., Bayraktar, E., Miskioglu, I., Robert, M.: Design and physical properties of multifunctional structural composites reinforced with nanoparticles for aeronautical applications. Adv. Mater. Process Technol. **3**(1), 33–44 (2016)
15. Ferreira, L.-M.-P., Bayraktar, E., Robert, M.-H.: Magnetic and electrical properties of aluminum matrix composite reinforced with magnetic nano iron oxide (Fe_3O_4). Adv. Mater. Process Technol. **2**(1), 165–173 (2016)
16. Katundi, D., Ferreira, L.-M., Bayraktar, E., Robert, M.-H.: Design and microstructural evolution, mechanical and physical properties of fine particles reinforced aluminum matrix composites. Adv. Mater. Process Technol. **2**(4), 566–577 (2016). https://doi.org/10.1080/2374068X.2016.1247244hal-01878664
17. Inamura, T., Fukui, Y., Hosoda, H., Wakashima, K., Miyazaki, S.: Mechanical properties of Ti–Nb biomedical shape memory alloys containing Ge or Ga. Mater. Sci. Eng. C. **25**, 426–432 (2005)

Check for updates

Chapter 6
Recycled Ti-Al-Cu Matrix Composites Reinforced with Silicon Whiskers and γ-Alumina (Al_2O_3) Fibres Through Sintering + Forging

Fabio Gatamorta, Ibrahim Miskioglu, Dhurata Katundi, and Emin Bayraktar

Abstract In this study, the microstructural formation and static/dynamic compression behaviour of the recycled Ti-Al-Cu matrix hybrid composites reinforced with silicon whiskers and alumina (Al_2O_3) fibres are examined. It is intended to be an alternative to traditional alloys/composites used in the aeronautical industry. These composites are generally produced using combined sintering + forging processes. The static and dynamic properties are evaluated in detail, taking into account the relevant scanning electron microscopy (SEM) microstructures, including the distribution of reinforcement elements.

Keywords Recycled composites · Silicon carbide · Fibre aluminum oxide · SEM-microstructure · Static and dynamic tests

6.1 Introduction

Titanium-aluminium-based (Ti-Al) composites with ceramic reinforcements are used mainly in aeronautical engineering applications as structural materials because of the combined mechanical properties (e.g., ductility and toughness), with the addition of high modulus and strength of the ceramics and/or ceramic fibres [1–12].

Even the new and expensive manufacturing methods, such as spark plasma sintering (SPS), of the special parts from these composites are thoroughly developed. The powder metallurgy, very often with the combined method (i.e., sintering + forging), are used for the sake of simplicity. In the present work, three special composite groups were designed in the frame of a common research project with Aeronautical Society in France.

The addition of ceramic fibres, together pure electrical copper to keep ductility and toughness, improve certain physico-chemistry parameters [8–18]. In fact, general defects (e.g., microcracks, microsegragations) and also agglomeration of the reinforcement particles in the grain boundaries disturb the the mechanical properties of these composites [2, 4, 6, 16, 18–22].

Recycled Ti-Al-Cu-based composites reinforced with silicon whiskers and alumina (Al_2O_3) fibres were processed through the combined method of sintering + forging. Mechanical and physical properties with different compositions have been compared.

6.2 Experimental Conditions

In the frame of the common research project, three different composites were designed from fresh scrap plus gas atomized Ti-60 wt.%Al intermetallic matrix with the reinforcement of pure electrolitic copper and silicon whiskers and alumina (Al_2O_3) fibres together. All the reinforcement powders of purity were at the level of 99.5%.

F. Gatamorta
UNICAMP, University of Campinas, FEM, Campinas, SP, Brazil
e-mail: fabiog@fem.unicamp.br

I. Miskioglu
ME-EM Department, Michigan Technical University, Houghton, USA
e-mail: imiski@mu.edu

D. Katundi · E. Bayraktar (✉)
School of Mechanical and Manufacturing Engineering, ISAE-SUPMECA-Paris, Paris, France
e-mail: dhurata.katundi@isae-supmeca.fr; emin.bayraktar@isae-supmeca.fr

V. Chalivendra, F. Gardea (eds.), *Mechanics of Composite, Hybrid & Multi-functional Materials, Volume 5*, Conference Proceedings of the Society for Experimental Mechanics Series, https://doi.org/10.1007/978-3-031-17445-2_6

Three types of composites were processed by using powder metallurgy-combined method of sintering + forging. Three composites were sintered at the same temperature (650 °C) followed by forging. Also certain specimens were only sintered to compare with the combined method. As a result, mechanical properties of the sintered forging composites were considerably improved.

In Table 6.1, the general compositions of the three composites are presented with the major and minor reinforcements. These compositions are actually under development to optimize the interface matrix-reinforcements for the research project that is ongoing.

The microstructure analyses have been carried out by use of a scanning electron microscope (SEM) with "EDS" chemical analyses. The density measurements of the specimens were carried out by using the Archimedes method. These values change between 2.85 and 3.28 g/cm^3. All the mechanical tests were conducted on a Zwick mechanical test device installed at the laboratory of ISAE-SUPMECA-Paris, France.

Three cylindrical specimens (H/D≥1.5) were tested under quasi-static and cyclic compression conditions according the DIN 50106 norm by using a constant test speed as 1 and 1,5 mm/min. Finally, low-velocity impact (drop-weight) tests were carried out with a drop tower to evaluate the response of the composites to dynamic loading. Microhardness ($HV_{0.1}$) tests were conducted on the polished and etched specimens. The mean microhardness values are presented in Table 6.2 with ±15–20% accuracy for the two manufacturing processing techniques, respectively.

6.3 Results and Discussions

Microstructure analyses starting from analyses of Ti-Al intermetallic and microstructure evaluation of the composite Ti-Al-Cu are presented in Fig. 6.1. EDS chemical analyses are also presented in this figure.

As usual, Ti_3Al and Cu_3Al intermetallic phases occurred with a solid chemical fusion during the sintering, Addition of pure electrolitic copper powder improved chemical bonding diffuision between matrix and the reinforcements; essentially, ceramic particles are firmly attached in the matrix without decohesion the with matrix. More detailed analyses should be necessary in case of the high amount of ceramic particles in the matrix. These results are limited under the laboratory conditions.

Static 3-point bending (3PB) test results are presented in Fig. 6.2. Bending stress evolution depends on the deformation carried out on the thin composite sheets; the hot forging process after sintering and also 3PB-test action done on the Zwick test machine are given in detail.

Static 3PB tests have been carried out for each composition here that were not given all the results. It seems that an acceptable ductility was observed, however; ductility increased evenly because of the higher amount of Ti-Al intermetallics added in the composites, J-II and J-III. As for mechanical behaviour under the static and cyclic compression solliciations, very large experimental data were obtaned from these mechanical tests for each composition [16–22]. Figures 6.3 and 6.4 give detailed information about these evaluations.

Figure 6.3 presents some of the test specimens prepared for static and cyclic compression tests and tests carried out on the Zwick test machine, and also a static compression test for the composite J-I only sintered at 600 °C. Figure 6.4 summarizes the mean values of the static and cyclic compression test results. It seems that there is not so much difference between the

Table 6.1 Compositions of the specimens (wt.%)

Specimen name	Ti-Al Fresh scrap	AA7075 Fresh scrap	Electrolitic copper, Cu	γ-Al_2O_3 fibre	SiC whiskers	hBN 0,5	GNP	Zn St
J-I	B	30	30	3	0,5	0,5	0,5	2
J-II	B	25	30	3	0,5	0,5	0,5	2
J-III	B	20	30	3	0,5	0,5	0,5	2

Table 6.2 Measurements of the microhardness values of the composite TiA-Cue-Fe_3O_4

Composition name	Microhardness values, $HV_{0,1}$
1st group (Sintered at 600 °C + Forged)	190 ± 10
1st group (Sintered only at 600 °C)	135 ± 02
2nd group (Sintered at 600 °C + Forged)	225 ± 05
3rd group (Sintered at 600 °C + Forged)	230 ± 15

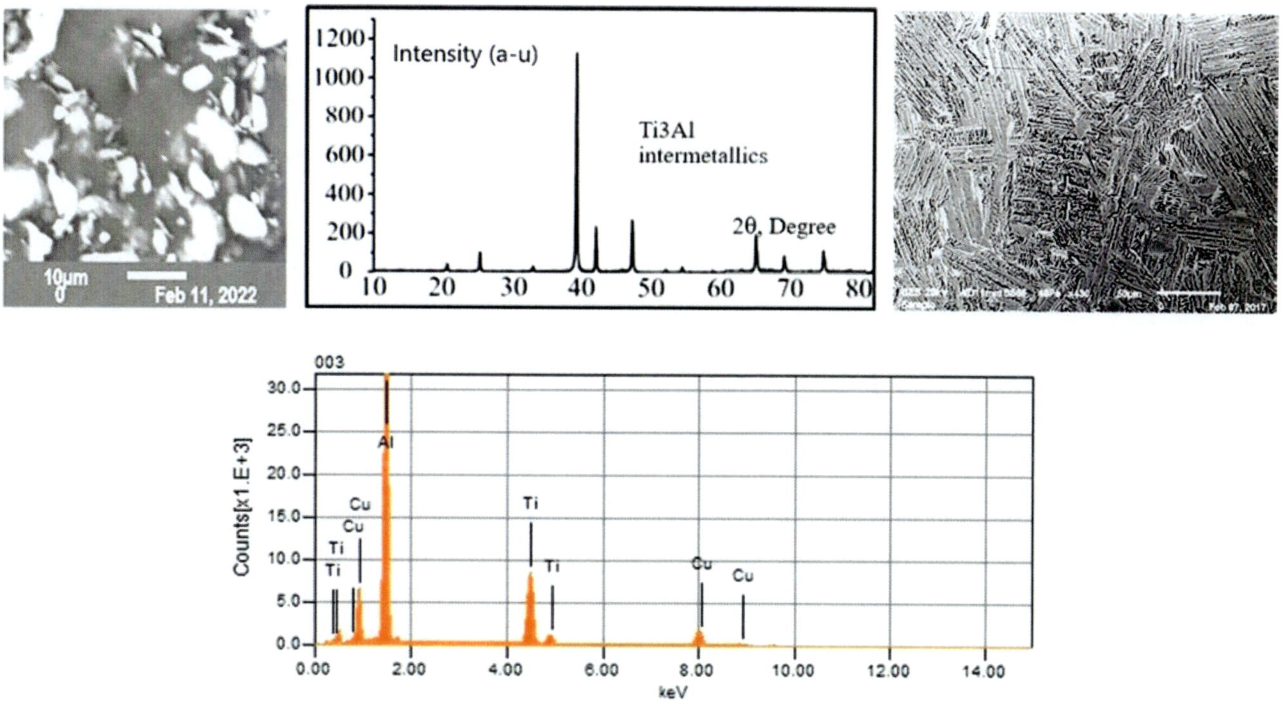

Fig. 6.1 Presentation of Ti-Al intermetallic powder taken from SEM and the microstructure evaluation of the composite Ti-Al-Cu; also XRD and EDS chemcial analyses of the composites designed here from gas-atomized fresh scraps

Fig. 6.2 Static 3PB test results: Bending test device on the Zwick test machine and finally evolution of bending stress as a function of the deformation

static compression test results of the three composites, ligthly increasing ultimate tensile stress (UTS) levels in J-II and J-III because of the increase of Ti-Al in these compositions.

Nevertheless, cyclic compression test (load–unload, stress–strain curves) results can give relatively backward stress if the load–unload stress evolution is not the same type. Similarly the Bauschinger effect is seem as a major reason for this evolution. Although the initial stress levels of the materials have similar magnitude, evolution of the load stresses are different. This effect is much higher than two (J-I, J-II) other composites and the final damage level (UTS).

Figure 6.5 shows typical SEM-fracture surfaces obtained from the damaged specimens solicited by cyclic-compression tests. Damage analyses, shown in Fig. 6.4, can be justified with these fracture surfaces. It means that the cumulative stresses can result in this type of damaged surfaces. Even this effect can be observed faintly on the fracture surface of the composite J-II.

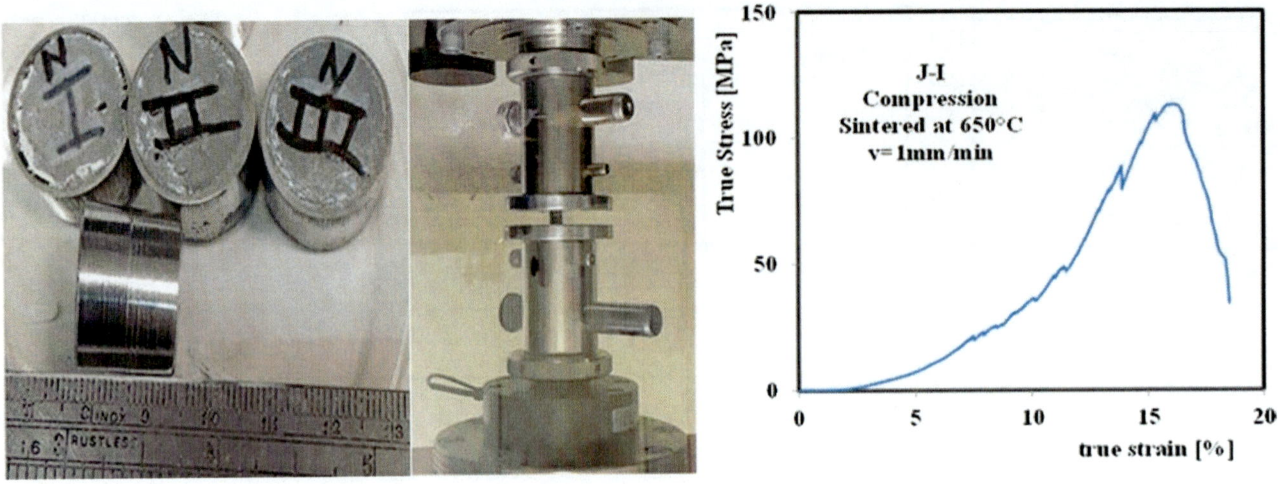

Fig. 6.3 Specimens prepared for static- and cyclic-compression tests and tests carried out on the Zwick test machine and static compression tesst for only the composite J-I, sintered at 600 °C

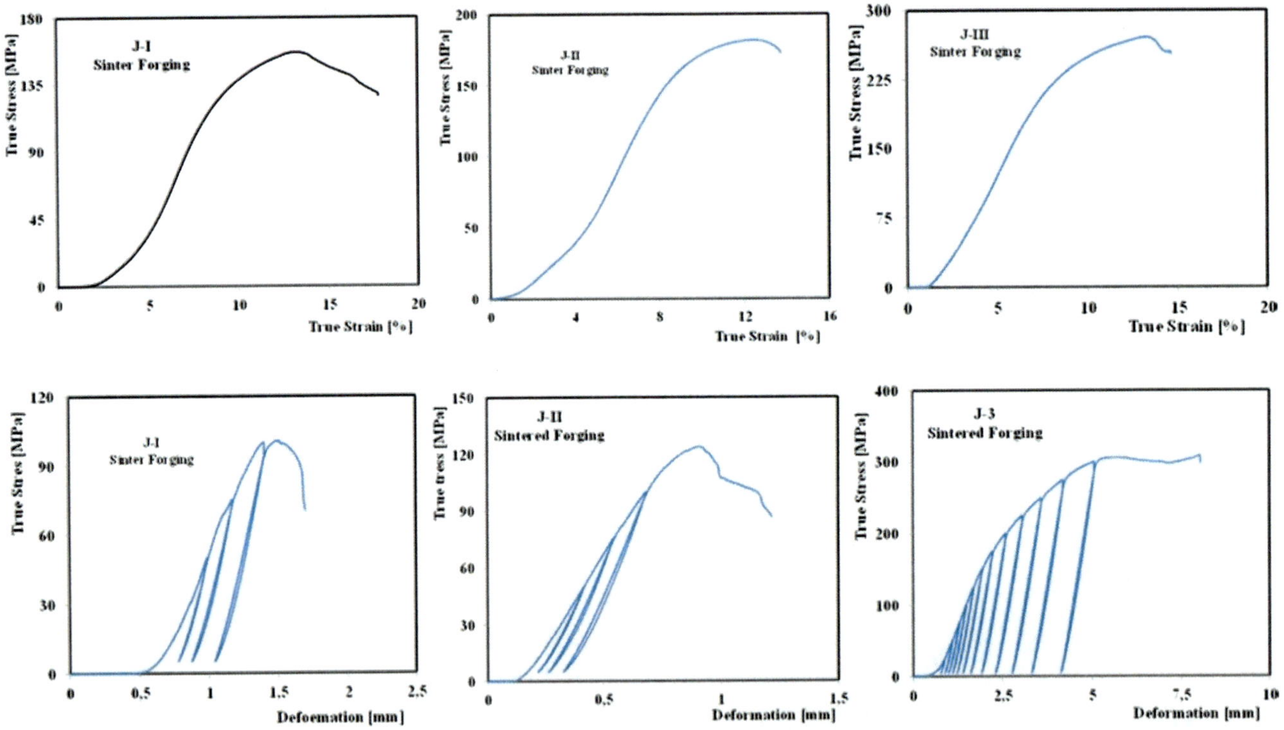

Fig. 6.4 Typical static- and cyclic- (load–unload curves) compresion tests for three compositions for the samples prepared from the composites through the combined method: sintering + forging

J-I. J-II. J-III.

Fig. 6.5 Typical SEM-fracture surfaces obtained from the damaged specimens solicited by cyclic-compression tests

Table 6.3 Average results for the energy absorption performance

Composite name	Impact energy, J	Absorbed energy, J	Energy absorption, %
J-I	11.5	7.340	82.13
J-II	11.5	8.263	84.44
J-III	11.5	9.775	85.38

Low-velocity impact (drop-weight) test results are presented in Table 6.3. This type of the test has been done by means of a drop tower to evaluate the response of the composites to dynamic loading. The force-displacement curves were presented for the compositions, and then the region under the load-displacement curve was calculated to determine the absorbed energy for each composition. Energy absorbed on impact and energy absorption performance are presented in Table 6.3.

6.4 Conclusions

In the frame of a common research project, a new aeronautical composite was designed in three different compositions, called intermetallic-based composites, Ti-Al-Cu. This composite was targeted to adapt easily for the new design of the parts used in a turbocompressor. Microstructural analysis has shown that a good bonding at the interface of matrix-reinforcement is essentially in the specimens manufactured with a combined process—sintering + forging—to produce a tough and sound microstructure. Cyclic loading–unloading behaviours of these specimens produced with sintering + forging can give certain behaviour similar to the Bauschinger effect that is seem as a major reason for this evolution.

Acknowledgements The authors acknowledge financial support from CNPq-National Council for Scientific and Technological Development (Brazil) and Program French Cathedra UNICAMP/French Consulate in São Paulo, Brazil. The third author, E. Bayraktar, extends personal thanks to the Department of FEM, University of UNICAMP, Campinas/SP-BRAZIL, and the French Consulate in Sao Paulo for partial financial support for this research project.

References

1. Torralba, J.M., Da Costa, C.E., Velasco, F.: Powder metallurgy-aluminium matrix composites, an overview. J. Mater. Process. Technol. **133**(1), 203–206 (2003)
2. Mihlyuzova, L., Enginsoy, H.M., Dontchev, D., Bayraktar, E.: Tailored behaviour of scrap copper matrix composites reinforced with zinc and aluminium: low cost shape memory structures. In: Mechanics of Composite, Hybrid and Multifunctional Materials, Vol. 6, pp. 27–34 (2020)
3. Kai, X.Z., Tian, K.L., Wang, C.M., Jiao, L., Chen, G., Zhao, Y.T.: Effects of ultrasonic vibration on the microstructure and tensile properties of the nano ZrB2/2024Al composites synthesized by direct melt reaction. J. Alloy Compd. **668**, 121–127 (2016)
4. Gatamorta, F., Enginsoy, H.M., Bayraktar, E., Miskioglu, I., Katundi, D.: Design of recycled alumix-123 based composites reinforced with γ-Al_2O_3 through combined method; sinter + forging. In: Mechanics of Composite, Hybrid and Multifunctional Materials, Vol. 6, pp. 9–18 (2020)
5. Koli, D.K., Agnihotri, G., Purohit, R.: Advanced aluminium matrix composites: the critical need of automotive and aerospace engineering fields. Mater Today Proc. **2**(4–5), 3032–3041 (2015)
6. Bayraktar, E., Gatamorta, F., Enginsoy, H.M., Polis, J.E., Miskioglu, I.: New design of composites from fresh scraps of niobium for tribological applications. In: Mechanics of Composite, Hybrid and Multifunctional Materials, Vol. 6, pp. 35–44 (2020)
7. Wang, P., Gammer, C., Brenne, F., Niendorf, T., Eckert, J., Scudino, S.: A heat treatable TiB2/Al-3.5Cu-1.5Mg-1Si composite fabricated by selective laser melting: microstructure, heat treatment and mechanical properties. Composites Part B. **147**, 162–168 (2018)
8. Enginsoy, H. M., Gatamorta, F., Bayraktar, E., Miskioglu, I., Larbi, A.: Design of copper and silicon carbide (SiC) reinforced recycled aluminium matrix composites through sintering + forging. In: Mechanics of Composite, Hybrid and Multifunctional Materials, Vol. 6, pp. 45–52 (2020)
9. Enginsoy, H.M., Bayraktar, E., Miskioglu, I., Katundi, D.: Manufacturing of "Ni-Ti" based composites from fresh scrap thin sheets reinforced with Nb and TiB2 through hot-forged bonding: sandwich structures. In: Mechanics of Composite, Hybrid and Multifunctional Materials, Vol. 6, pp. 53–60 (2020)
10. Chen, L., Lin, J., Xiangjun, X., Li, C., Yong, X., Liang, Y.: Microstructure refinement via martensitic transformation in TiAl alloys. J. Alloys Compd. **741**, 1175–1182 (2018)
11. Bayraktar, E., Katundi, D., Gatamorta, F., Miskioglu, I., Murat Enginsoy, H.: Design of recycled thin sheet "Ti-Al" based composites reinforced with AA1050, boron, TiB2, TiC, and B4C through hot-forged bonding. In: Mechanics of Composite, Hybrid and Multifunctional Materials, Vol. 6, pp. 65–72 (2020)
12. Enginsoy, H.M., Bayraktar, E., Miskioglu, I., Gatamorta, F., Katundi, D.: Design of copper and γ-alumina reinforced recycled aluminium matrix composites through sintering + forging. In: Mechanics of Composite, Hybrid and Multifunctional Materials, Vol. 6, pp. 73–80 (2020)

13. Enginsoy, H.M., Miskioglu, I., Bayraktar, E., Katundi, D.: Static and dynamic behaviour of recycled AA7075 based composites reinforced with ZrO2-γ-Al203 fibre and SiC. In: Mechanics of Composite, Hybrid and Multifunctional Materials, Fracture, Fatigue, Failure and Damage Evolution, Vol. 3, pp. 29–34 (2021)
14. Gatamorta, F., Enginsoy, H.M., Bayraktar, E., Miskioglu, I.: Recycled "Al431 + A1050" based composites reinforced with "TiC" ceramic powders for aeronautical applications. In: Mechanics of Composite, Hybrid and Multifunctional Materials, Fracture, Fatigue, Failure and Damage Evolution, Vol. 3, pp. 35–40 (2021)
15. Mihlyuzova, L., Enginsoy, H.M., Stanislav, S., Dontchev, D., Bayraktar, E.: Experimental and finite element study of recycled aluminium (AA7075) matrix composites reinforced of TiC/MoS2/γ-Al2O3Fibre/Nb2Al. In: Mechanics of Composite, Hybrid and Multifunctional Materials, Fracture, Fatigue, Failure and Damage Evolution, Vol. 3, pp. 41–48 (2021)
16. Mihlyuzova, L., Enginsoy, H.-M., Bayraktar, E., Slavov, S., Dontchev, D.: Tailored behaviour of scrap copper matrix composites reinforced with "Zn-Ni-Al", low cost shape memory structures. In: Mechanics of Composite, Hybrid and Multifunctional Materials, Fracture, Fatigue, Failure and Damage Evolution, Vol. 3, pp. 49–54 (2021)
17. Munoz-Morris, M.A., Rexach, J.I., Lieblich, M.: Comparative study of Al-TiAl composites with different intermetallic volume fractions and particle sizes. Intermetallics. **13**, 141–149 (2005)
18. Zambelis, Enginsoy, H.-M. Bayraktar, E., Larbi, A., Katundi, D.: Mechanical properties of recycled rubber modified epoxy resin based composites for aircraft auxiliary structures. In: Mechanics of Composite, Hybrid and Multifunctional Materials, Fracture, Fatigue, Failure and Damage Evolution, Vol. 3, pp. 55–64 (2021)
19. Enginsoy, H.M., Bayraktar, E., Katundi, D., Larbi, A.: Manufacturing of scrap thin sheet "Ni-Ti-Al" based composites reinforced with Nb2Al through hot-forged bonding: sandwich structures. In: Mechanics of Composite, Hybrid and Multifunctional Materials, Fracture, Fatigue, Failure and Damage Evolution, Vol. 3, pp. 65–70 (2021)
20. Enginsoy, H.M., Bayraktar, E., Gatamorta, F., Katundi, D., Miskioglu, I.: Recycled niobium-aluminium (Nb_2Al) intermetallics based composite design: an experimental and numerical approach for toughening mechanism. In: Mechanics of Composite, Hybrid and Multifunctional Materials, Fracture, Fatigue, Failure and Damage Evolution, Vol. 3, pp. 71–80 (2021)
21. Pereira, V.L., Almeida, I.S., Neto, S.A., Bayraktar, E., Ferreira, L.P., Concretes, A.: Study of concrete performance with addition of copper tailings reinforced and steel fiber, mechanics of composite, hybrid and multifunctional materials, fracture, fatigue, failure and damage. Evolution. **3**, 89–92 (2021)
22. Bower, A.F.: Cyclic hardening properties of hard-drawn copper and rail steel. J. Mech. Phys. Solids. **37**(4), 455–470 (1989)

Chapter 7
Development of Ni-Al/Nb_2Al/ZrO_2-Based Composites for Aircraft Engine Applications Produced by a Combined Method: Sintering + Forging

Utku Kaftancıoglu, Georges Zambelis, Fabio Gatamorta, Ibrahim Miskioglu, and Emin Bayraktar

Abstract In this study, the microstructural formation and static/dynamic compression behaviour of the recycled Ni-Al/Nb_2Al/ZrO_2 matrix-hybrid composites reinforced with Nb and ZrO_2 will be studied. It is intended to be an alternative to traditional alloys/composites used in the aeronautical industry. These composites are generally produced by using combined sintering + forging processes. The static and dynamic properties will be evaluated in detail, considering the relevant scanning electron microscopy (SEM) microstructures, including the distribution of reinforcement elements.

Keywords NiAl-recycled composites · Nb_2Al · ZrO_2 · SEM microstructure · Static and dynamic tests

7.1 Introduction

In recent aeronautical applications, nickel aluminide (Ni-Al) matrix composites are considered as a better alternative to Ni-based super alloys because of their low density with great oxidation resistance and thermal conductivity. To achieve high-strength responses at operating temperatures of aeronautical applications (600–850 °C), compounds (e.g., intermetallic Nb_2Al and Zr_2O_3) are preferred for reinforcement materials [1, 2]. Essentially, Ni-Al metal matrix composites provide high toughness and strength at elevated temperatures that can match the super alloys used in the industry while keeping the weight of the aircraft components as small as possible. For the past decade, the focus of several studies has been on the improvement of metal-matrix composites by using particle reinforcements on micro and nano scales.

Development of these hybrid composites unlocks new possibilities in automotive industry as well as aeronautical and aerospace applications in terms of material characteristics. This is because mechanical properties of metal-matrix-hybrid composites can be modified according to the specific needs of an application area [3–7]. Essentially, addition of nanoparticles along with fine ceramic particles improves the mechanical properties of the composites much more efficiently; as a result, it is possible to obtain the required strength and toughness along with the ductility of the composites [8].

U. Kaftancıoglu
Department of Mechanical Engineering, Atilim University, Ankara, Turkey

School of Mechanical and Manufacturing Engineering, ISAE-SUPMECA-Paris, Paris, France
e-mail: kaftancioglu.utku@student.atilim.edu.tr

G. Zambelis
Composites Research and Development Centre, Airbus-Helicopters, Paris, France
e-mail: georges.zambelis@airbus.com

F. Gatamorta
UNICAMP-FEMUniversity of Campinas, Campinas, SP, Brazil
e-mail: fabiog@fem.unicamp.br

I. Miskioglu
ME-EM Department, Michigan Technological University, MI, USA
e-mail: imiski@mtu.edu

E. Bayraktar (✉)
School of Mechanical and Manufacturing Engineering, ISAE-SUPMECA-Paris, Paris, France
e-mail: emin.bayraktar@isae-supmeca.fr

V. Chalivendra, F. Gardea (eds.), *Mechanics of Composite, Hybrid & Multi-functional Materials, Volume 5*, Conference Proceedings of the Society for Experimental Mechanics Series, https://doi.org/10.1007/978-3-031-17445-2_7

In aeronautical applications, mostly near net shape-forming processes of powder metallurgy such as hot isostatic pressing (HIP) + hot isothermal forging (HIF) and HIP + hot extrusion + HIF are preferred for manufacturing of lightweight metal- or hybrid-matrix composites [9, 10]. Because most of these routes include expensive and inefficient processes (e.g., HIP), a combined powder metallurgy route is used with sintering + forging of powders obtained from scrap metallic parts. With this combined method, the micro-defects of the scrap parts are virtually eliminated; therefore, it is possible to obtain lightweight metal matrix composites approaching the theoretical densities of the composites [11–13].

This work was designed to study the compression behaviour of recycled Ni-Al matrix-hybrid composites reinforced with Nb_2Al and ZrO_2 under static and dynamic conditions for various particle reinforcement compositions. The matrix-reinforcement interaction was studied by means of scanning electron microscopy (SEM) along with the mechanical properties of the microstructure.

7.2 Experimental Conditions

In the frame of the common research project, five different composites were designed from fresh scrap + gas-atomized Ni-60wt.%Al intermetallic matrix with the reinforcement of Nb_2Al and ZrO_2 (i.e., a mixture of Zr + ZrO_2) together. All the reinforcement powders of purity were at the level of 99.5%. Three types of composites were processed by using a powder metallurgy-combined method: sintering + forging. These composites were sintered at the same temperature (650 °C) followed by forging. Also certain samples of the first group were only sintered to compare with the combined method. Accordingly, mechanical properties of the sintering + forging composites were considerably improved.

In Table 7.1, the general compositions of the three composites are presented with the major and minor reinforcements. Developments of these composites are being done to optimize mechanical properties and improve the interface matrix-reinforcements for the ongoing common research project.

Figure 7.1 explains typical Ni-Al binary-phase boundary conditions. The composite prepared here (i.e., Ni-60wt.%Al intermetallic-based) has been validated with EDS chemical and XRD analyses. As shown in the diagram, a major amount of the phase is "Ni-Al" and the minor phase in the composite is Al_3Ni_2. In fact, doping and mixing of the reinforcement to the matrix should improve through the laboratory conditions.

The microstructural analyses were carried out by SEM with "EDS" chemical analyses. The density measurements of the specimens were taken by using the Archimedes method. These values change between 6.95 and 8.15 g/cm^3. All the mechanical tests were carried out on a Zwick mechanical test device installed at the laboratory of ISAE-SUPMECA-Paris, France.

Three prepared cylindrical specimens (H/D≥1.5) were tested under quasi-static- and cyclic-compression conditions according the EN-standard = DIN 50106 by using a constant test speed as 1 and 1,5 mm/min. Additionally, low-velocity impact (drop-weight) tests were carried out with a drop tower to evaluate the response of the composites to dynamic loading. Microhardness ($HV_{0.1}$) tests were performed on the polished and etched specimens. The mean microhardness values are presented in Table 7.2 with ±5–10% accuracy for the two manufacturing processing techniques, respectively.

Table 7.1 General Composition of the Composite UT-I (Ni-Al/Nb_2Al-ZrO_2)

Composition (wt.%)	Ni-Al (60/40) (atomized)	Nb_2Al	Zr	ZrO_2	Electrolytic Cu	Zn-stearate
UT-Ia	B	10	3	7	2	2
UT-Ib	B	15	3	7	2	2
UT-Ic	B	10	3	12	2	2

Table 7.2 Measurements of the Microhardness Values of the Composite (Ni-Al/Nb_2Al-ZrO_2)

Composition name	$HV_{0,1}$
UT-Ia (Sintered at 650 °C + Forged)	185 ± 10
UT-Ib (Sintered only at 650 °C)	230 ± 05
UT-Ic (Sintered at 650 °C + Forged)	240 ± 10

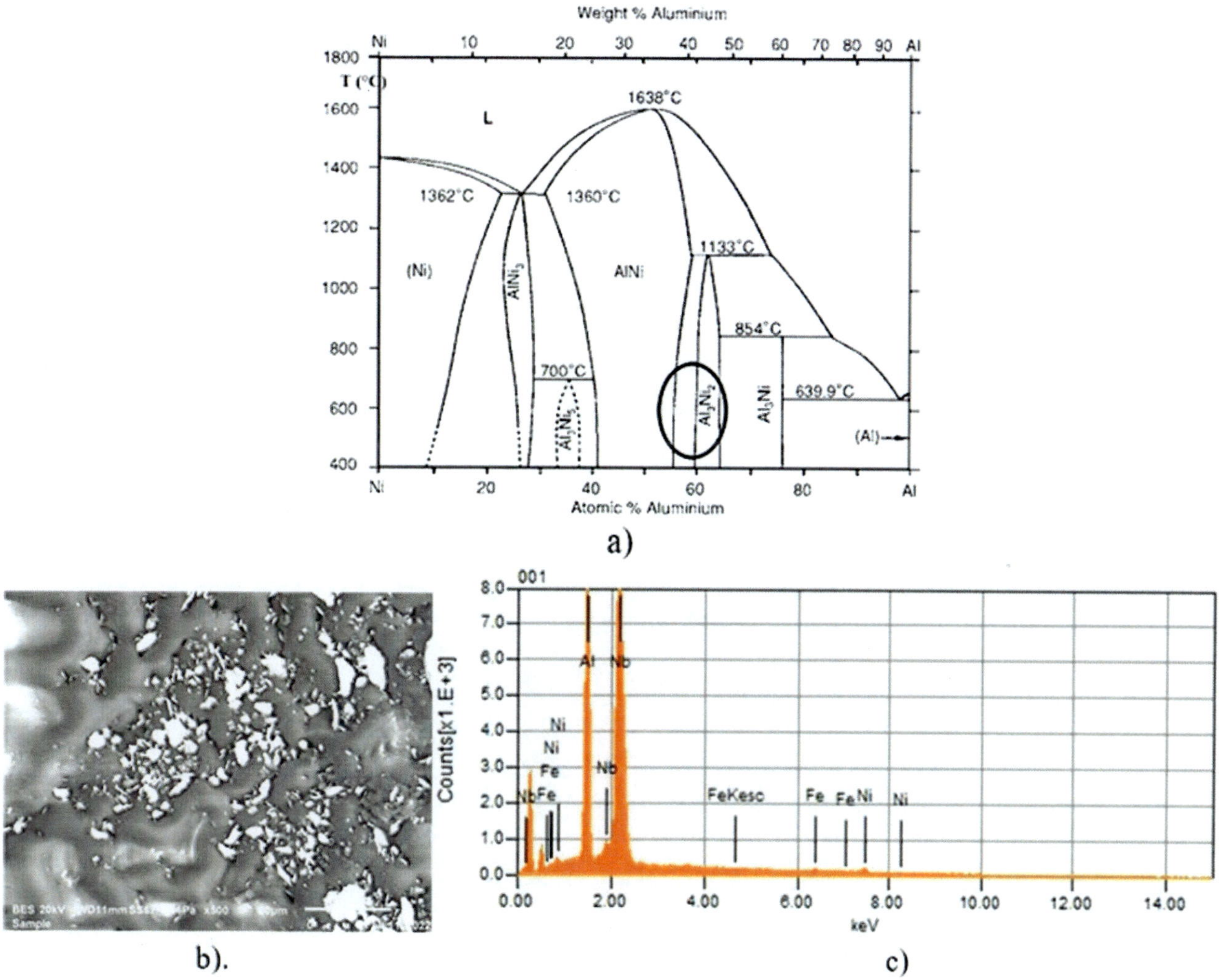

Fig. 7.1 Ni-Al binary one-phase boundary: (**a**) Final stage of the composite after doping and mixing of the reinforcement with matrix Ni-60wt.%Al intermetallic; (**b**) EDS chemical analyses

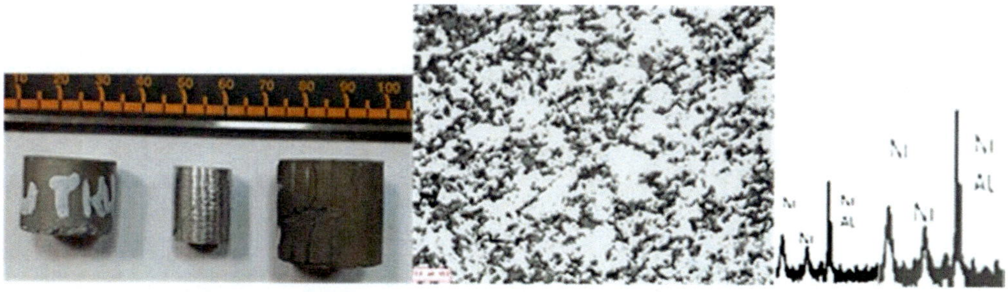

Fig. 7.2 Ni-Al intermetallic-based composite powder after 4 h milling; also XRD and EDS chemical analyses with microstructure of the composites (UT-I) designed here from gas-atomized fresh scraps

7.3 Results and Discussions

Microstructure analyses starting from analyses of Ni-Al intermetallic and powder evaluation of the final composite (Ni-Al/Nb_2Al-ZrO_2) are presented in Fig. 7.1. "EDS" chemical analysis is also presented in that figure.

As usual, intermetallic phases occur with a solid chemical fusion during the sintering. Addition of pure hard intermetallic Nb_2Al and very hard ZrO_2 ceramic powder increased an effective chemical bonding diffusion between the matrix and the

Fig. 7.3 Hot forging testing device after sintering the UT-I specimen at 650 °C (*left*); compression test device on the Zwick test machine (*centre*) and damaged specimen (45°cracks max. shear stress) after compression test (*right*)

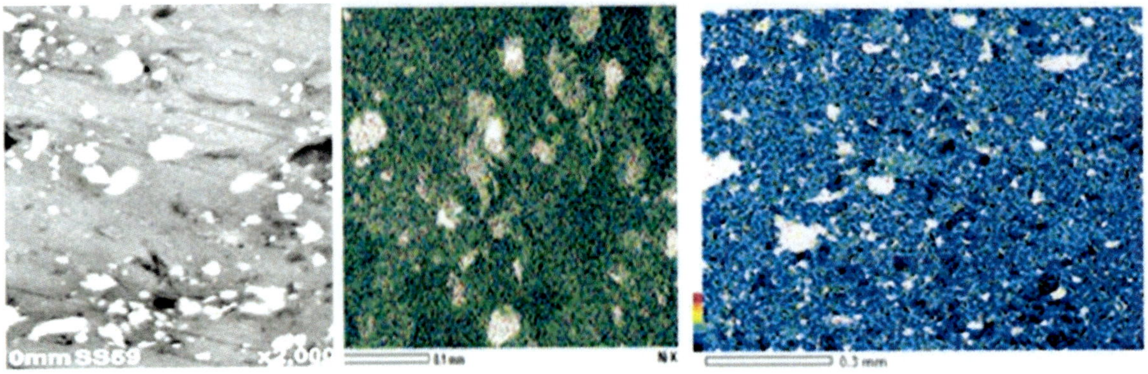

UT-Ia

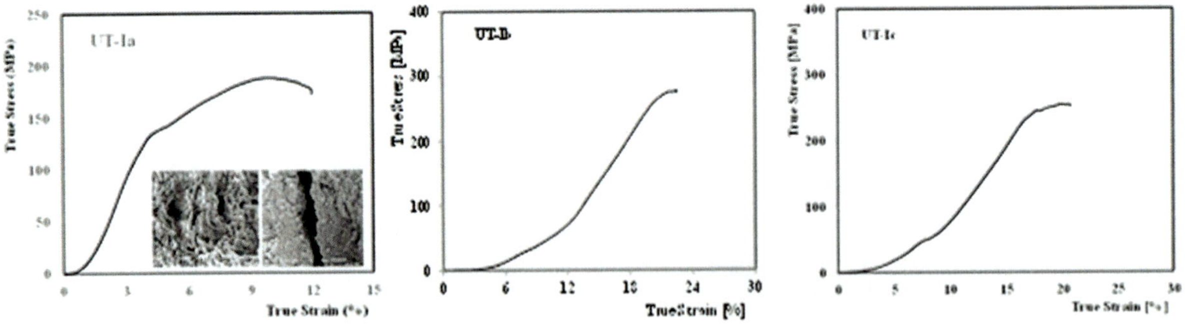

UT-Ib. UT-Ic

COMPOSITE	UTS (MPa)
UT-Ia	187
UT-Ib	275
UT-Ic	255

Fig. 7.4 Microstructural evaluation with elementary mapping analyses, fracture surface and crack appearance at the surface for the composition UT-Ia; also static-compression test graphs and max. strength values for UT-Ia, b and c compositions

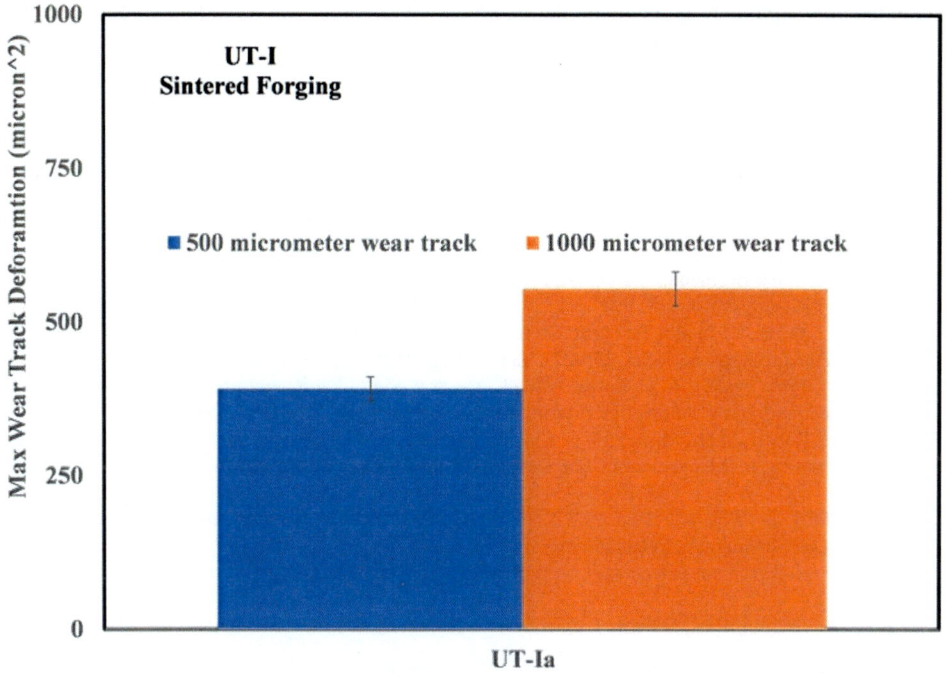

Fig. 7.5 Maximum wear track deformation measured as the area between the initial profile and the final (residual) profile of the wear tracks for 500 and 1000 μm long-wear paths for the UT-Ia (Ni-Al/Nb_2Al-ZrO_2) specimens

reinforcements attached in it without a decohesion with the matrix. More detailed analyses would be needed in the case of the high quantity of ceramic particles in the matrix. These results were limited under the laboratory conditions.

Figure 7.2 presents test results for specimens under static compression and the general microstructure of the composite UT-Ia with XRD phase analyses. Depending on the milling time, the homogenous distribution of the reinforcements in the matrix increased. Hot forging testing device, after sintering the specimen UT-I at 650 °C, and the compression test device on the Zwick test machine and also damaged specimens can be seen in Fig. 7.3. On the damaged specimen, maximum shear stress (cracks 45°) is shown very clearly after a compression test. Virtually all the specimens from three compositions showed the same behaviour [14–17].

Figure 7.4 shows typical microstructural evaluation with elementary mapping analyses for the distribution of Ni-Al, Nb_2Al and ZrO_2 fracture surfaces and crack appearance at the surface for only the composition UT-Ia. It seems that the distribution of each element in the structure was homogenous with few agglomerations. This has a real effect on the mechanical behavior and resistance of the composites under static solicitation.

In the same figure, the static-compression tests results are presented for three compositions. Also maximum strength values (UTS) were given in a Table 7.2 for these compositions of UT-Ia, b and c. It seems that the influence of Nb_2Al intermetallic and also ZrO_2 ceramic oxide reinforcements are considerable regarding to the first composition, UT-Ia.

In case of time-dependent behaviour by means of nanoindentation, some of the wear tests displayed in Fig. 7.5 shows the maximum wear track deformation measured as the area between the initial profile and the final (residual) profile of the wear tracks for 500 and 1000 μm long-wear paths for the UT-Ia (Ni-Al/Nb_2Al-ZrO_2) specimens [14–20]. These are the averages of the ten wear tests for obtaining a mean value. All the observations on the tested specimens (UT-Ia) indicate that the prepared samples resulted the need for a special surface treatment to increase the resistance of the composite against wear; essentially, this was the case for the parts used at temperatures beyond 400 °C.

7.4 Conclusions

In the frame of a common research project, a new aeronautical composite was designed in three different compositions, called intermetallic-based composites, Ni-Al/Nb_2Al-ZrO_2. These composites were intended to adjust simply for the parts used in the aeronautical area—for example, application of turbo compressor, engine parts and so on. Microstructural analysis has shown that there is a goodquality bonding at the interface of matrix-reinforcement in the specimens manufactured with the combined process called sintering + forging. This structure created tough and sound behaviour of the parts under mechanical solicitations.

Acknowledgements All the mechanical tests have been conducted on the Zwick mechanical test device installed at the laboratory of ISAE-SUPMECA-Paris, France; we acknowledge their technical help. We acknowledge the French Aeronautical Society, Airbus-Helicopters and Dr. G. Zambelis for his support for this project. The authors acknowledge Prof. I. Miskioglu from Michigan Technical University, MI-USA, for the nanoindentation tests for the wear behaviour of the composite studied in this project.

References

1. Lu, Y.X., Tao, C.H., Yang, D.Z.: Improvement in mechanical properties of Nial matrix composites fabricated by reaction compocasting. Scr. Mater. **35**(10), 1243–1246 (1996). https://doi.org/10.1016/1359-6462(96)00276-x
2. Gu, D., Wang, Z., Shen, Y., Li, Q., Li, Y.: In-situ Tic particle reinforced Ti–Al matrix composites: powder preparation by mechanical alloying and selective laser melting behavior. Appl. Surf. Sci. **255**(22), 9230–9240 (2009). https://doi.org/10.1016/j.apsusc.2009.07.008
3. Popov, V.A., Shelekhov, E.V., Prosviryakov, A.S., Kotov, A.D., Khomutov, M.G.: Particulate metal matrix composites development on the basisof in situ synthesis of TiC reinforcing nanoparticles during mechanical alloying. J. Alloys Compd. **707**, 365–370 (2017)
4. Ferreira, L.F.P., Bayraktar, E., Robert, M.H., Miskioglu, I.: Chapter 17: particles reinforced scrap aluminium based composites by combined processing sintering + thixoforming. In: Mechanics of Composite and Multi-functional Materials, Vol. 7, pp. 145–152. Springer, Berlin (2016). ISBN:978-3-319-41766-0
5. Lekatou, A., Karantzalis, A.E., Evangelou, A., Gousia, V., Kaptay, G., Gácsi, Z., et al.: Aluminum reinforced by WC and TiC nanoparticles (ex-situ) and aluminide particles (in-situ): Microstructure, wear and corrosion behaviour. Mater. Design. **65**, 1121–1135 (2015)
6. Bagheri, G.A.: The effect of reinforcement percentages on properties of copper matrix composites reinforced with TiC particles. J. Alloys Compd. **676**, 120–126 (2016)
7. Rudianto, H., Joo, J.K., Sun, Y.S., Jin, K.Y., Dlouhy, I.: Mechanical properties of sintered of Al-5.5Zn-2.5Mg-0.5Cu PM alloy. Mater. Sci. Forum. **794–796**, 501–507 (2014)
8. Gatamorta, F., Miskioglu, I., Bayraktar, E., Melo, M. L.: Recycling of aluminium-431 by high energy milling reinforced with Tic-Mo-Cu for new composites in connection applications. In: Mechanics of Composite and Multi-functional Materials, Vol. 5, pp. 41–46 (2019). https://doi.org/10.1007/978-3-030-30028-9_6
9. Singh, S., Gupta, D., Jain, V., Sharma, A.K.: Microwave processing of materials and applications in manufacturing industries: a review. Mater. Manuf. Process. **30**, 1–29 (2015). https://doi.org/10.1080/10426914.2014.952028
10. Teer, D.G., Hampshire, J., Fox, V., Bellido-Gonzalez, V.: The tribological properties of MoS2/metal composite coatings deposited by closed field magnetron sputtering. Surf. Coat. Technol. **94**, 572–577 (1997)
11. Ferreira, L.M.P., Robert, M.H., Bayraktar, E., Zaimova, D.: New design of aluminium based composites through combined method of powder metallurgy and thixoforming. Adv. Mater. Res. **939**(1), 68–75 (2014)
12. Srivyasa, P.D., Charoo, M.S.: Role of fabrication route on the mechanical and tribological behavior of aluminium metal matrix composites – a review. Mater. Today Proc. **5**, 20054–20069 (2018)
13. Ferreira, L.F.P., Bayraktar, E., Robert, M.H., Miskioglu, I.: Chapter 17: particles reinforced scrap aluminium based composites by combined processing sintering + thixoforming. In: Mechanics of Composite and Multi-functional Materials, pp. 145–152. Springer, Berlin (2016). ISBN: 978-3-319-41766-0
14. Mihlyuzova, L., Enginsoy, H.-M., Bayraktar, E., Slavov, S., Dontchev, D.: Tailored behaviour of scrap copper matrix composites reinforced with "Zn-Ni-Al", low cost shape memory structures. In: Mechanics of Composite, Hybrid and Multifunctional Materials, Fracture, Fatigue, Failure and Damage Evolution, Vol. 3, pp. 49–54 (2021)
15. Mihlyuzova, L., Enginsoy, H.M., Dontchev, D., Bayraktar, E.: Tailored behaviour of scrap copper matrix composites reinforced with zinc and aluminium: low cost shape memory structures. In: Mechanics of Composite, Hybrid and Multifunctional Materials, Vol. 6, pp. 27–34 (2020)
16. Kai, X.Z., Tian, K.L., Wang, C.M., Jiao, L., Chen, G., Zhao, Y.T.: Effects of ultrasonic vibration on the microstructure and tensile properties of the nano ZrB2/2024Al composites synthesized by direct melt reaction. J. Alloy Compd. **668**, 121–127 (2016)
17. Gatamorta, F., Enginsoy, H.M., Bayraktar, E., Miskioglu, I., Katundi, D.: Design of recycled alumix-123 based composites reinforced with γ-Al_2O_3 through combined method; sinter + forging. In: Mechanics of Composite, Hybrid and Multifunctional Materials, Vol. 6, pp. 9–18 (2020)
18. Koli, D.K., Agnihotri, G., Purohit, R.: Advanced aluminium matrix composites: the critical need of automotive and aerospace engineering fields. Mater. Today Proc. **2**(4–5), 3032–3041 (2015)
19. Bayraktar, E., Gatamorta, F., Enginsoy, H.M., Polis, J.E., Miskioglu, I.: New design of composites from fresh scraps of niobium for tribological applications. In: Mechanics of Composite, Hybrid and Multifunctional Materials, Vol. 6, pp. 35–44 (2020)
20. Wang, P., Gammer, C., Brenne, F., Niendorf, T., Eckert, J., Scudino, S.: A heat treatable TiB2/Al-3.5Cu-1.5Mg-1Si composite fabricated by selective laser melting: microstructure, heat treatment and mechanical properties. Composit. Part B. **147**, 162–168 (2018)

Chapter 8
Out-of-Plane Restraint Under Tension In-Plane Loading

Dawn C. Jegley, W. Allen Waters Jr., and Alana Zahn Cardona

Abstract The development and description of a test fixture designed to restrict out-of-plane motion in the center region of tension-loaded carbon-epoxy panels are presented herein. The test fixture was used to impose displacement conditions consistent with those in an analytical tool being developed to predict panel behavior in the vicinity of a central notch, which included the assumption that there would be no out-of-plane displacement and no buckling. However, pretest analysis using finite element models for a panel loaded without any out-of-plane restraint in the region of the notch indicated that with an imperfection magnitude equal to 20% of the thickness of the thin-skin panel, an unacceptable amount of out-of-plane deformation would occur as Poisson effects induced compression loads in the region of the notch. Therefore, to validate this tool in a test program, a restraint fixture which would suppress out-of-plane motion was required. The panel could not be encased in restraining plates because instrumentation and visibility were required in the vicinity of the notch. Therefore, a fixture that would restrict out-of-plane motion while still allowing access to the surface of the panel at the notch edges for instrumentation and line-of-sight access for cameras was required. To satisfy these requirements, a fixture was designed to restrict only out-of-plane motion near the center of the notch. Two 1.78-m-long test panels were loaded in tension to failure using this fixture. Out-of-plane deformations were not directly measured during testing, so back-to-back strain gages were used to obtain an indication of buckling. Strain results indicated that the restraint fixture performed as designed and buckling did not occur.

Keywords Transverse buckling · Fixtures · Carbon-epoxy · Notches · Out-of-plane restraints

8.1 Introduction

The goal of performing a mechanical test must be defined prior to designing the test arrangement. For example, an experiment may be intended to simulate a real-world condition in a laboratory environment, or be designed to enforce a simple and easily modeled condition to support the development of an analytical tool. Using the appropriate test fixturing can be a critical part of conducting a successful experiment on any structure. Even relatively simple experiments can require complex restraints or load-introduction systems to ensure the desired conditions are imposed on the test article.

The purpose of the test program supporting the tests described herein was to support the validation of a non-finite-element prediction methodology that could be used as a preliminary assessment and design tool to predict the behavior of carbon-epoxy stiffened panels with large notches. This validation effort required analysis and testing with tensile and compressive loads applied to two sizes of unstiffened panels as well as two stiffened panels of the same size that were larger than the unstiffened panels. The testing of some of the unstiffened test articles is described in Refs. [1, 2].

The tool being developed contained the assumptions of uniform tension or compression load introduction, no buckling, and no other out-of-plane displacement. An earlier version of this analytical tool is discussed in Ref. [3]. The need for validation of the preliminary-design rapid tool dictated the test-article geometry, stacking sequence, load, and boundary conditions.

D. C. Jegley (✉) · A. Z. Cardona
NASA Langley Research Center, Hampton, VA, USA
e-mail: dawn.c.jegley@nasa.gov; alana.cardona@nasa.gov

W. Allen Waters Jr.
Analytical Mechanics Associates, Inc., Hampton, VA, USA
e-mail: william.a.waters@nasa.gov

V. Chalivendra, F. Gardea (eds.), *Mechanics of Composite, Hybrid & Multi-functional Materials, Volume 5*, Conference Proceedings of the Society for Experimental Mechanics Series, https://doi.org/10.1007/978-3-031-17445-2_8

The validation effort involving the larger of the unstiffened panels loaded in uniaxial tension is discussed in this document. The testing of two carbon-epoxy panels with a central lateral notch was used to obtain the desired information for validation of the methodology for tensile loading.

8.2 Panel Description

For validation of the rapid tool, a carbon-epoxy laminate with stacking sequence $[+45/90/-45/0_2/+45/90/-45/0]_S$, containing 18 plies with a total thickness of approximately 0.33 cm, was chosen. Tension-loaded panels were 1.78 m long, 0.51 m wide, and contained a test section length of 1.52 m. Each panel contained a mid-length lateral notch which was 10.16 cm long with 0.635-cm-diameter circular ends. The notches widened in the center to 0.953 cm. This notch shape was chosen to be the same as the notch in similar compression-loaded panels (not discussed herein) to avoid contact of the notch edges during loading. The 0° fiber orientation was the loading direction. A sketch of the panel is shown in Fig. 8.1a, and a photograph of one of the test panels is shown in Fig. 8.1b.

8.3 Experiment Design

Prior to panel fabrication, an analysis was conducted using the finite element code Abaqus [4] to support selection of the test machine, to enable design of the load introduction interfaces, to develop the instrumentation plan, and to determine the need for a restraint system to prevent out-of-plane motion. Since the focus of the analytical tool was the behavior of the panel in the region of the notch, the behavior of the panel away from the notch was considered less important to the success of the test. An important assumption in the analytical tool was that no out-of-plane deformation would take place. Therefore, the pretest finite element analysis included an evaluation of the out-of-plane deformation that would occur in the region of the notch.

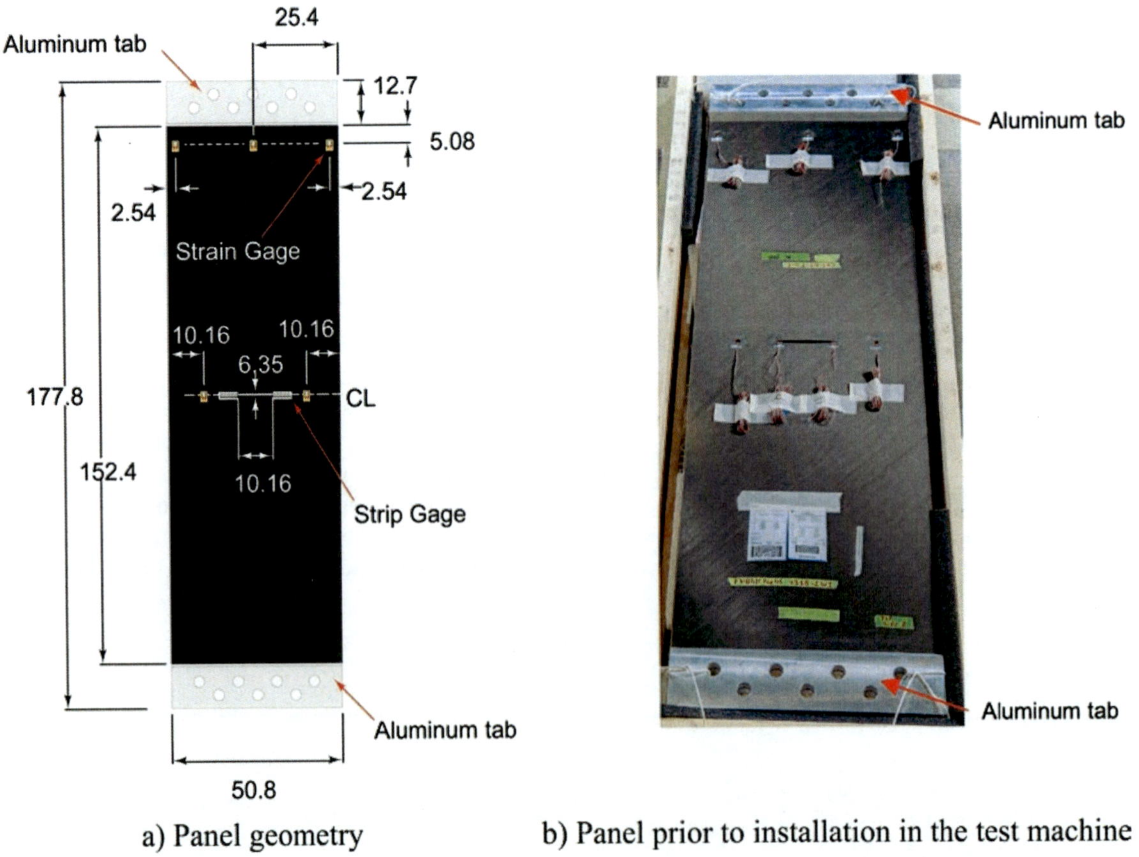

Fig. 8.1 Geometry of the tension panels with dimensions in cm. (**a**) Panel geometry (**b**) Panel prior to installation in the test machine

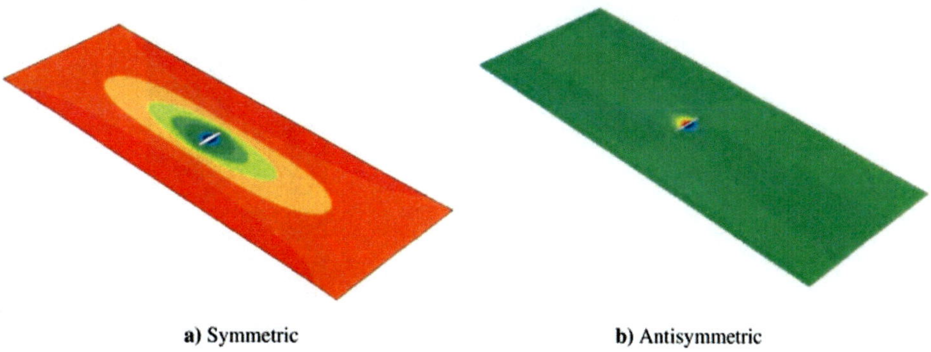

Fig. 8.2 Assumed imperfection shapes of geometrically-imperfect panel. (**a**) Symmetric (**b**) Antisymmetric

Geometric imperfections in thin-walled structures can have a significant influence on panel behavior [5]. These imperfections are critically important with respect to the buckling behavior of compression-loaded panels. In the case of the tension-loaded panel described herein, the central notch opened the possibility that the notch region might deform out-of-plane due to Poisson effects and therefore, geometric imperfections were assessed for their influence. In order to account for potential geometric imperfections, symmetric and antisymmetric deformation shapes were considered. The symmetric and antisymmetric imperfection patterns shown in Fig. 8.2 were investigated. This pretest analysis indicated that the antisymmetric mode was the more critical mode so only that mode is discussed herein.

An imperfection magnitude equal to 20% of the panel thickness for the antisymmetric imperfection case with out-of-plane restrictions only at the loaded edges of the panel resulted in an induced deformation at the notch tip of up to 18% of the panel thickness. This out-of-plane deformation at the notch tip of 0.0597 cm was deemed inappropriate for the validation effort, so a restraint fixture that would suppress out-of-plane motion was required. The test program required instrumentation and visibility in the region of the notch, so the panel could not be encased in supporting plates. A restraint system that would prevent out-of-plane motion, but would allow access to the surface of the panel near the notch edges for instrumentation and line-of-sight access for cameras was required.

Further analysis indicated that, if restraints to out-of-plane motion were added to the nodes mid-width as shown in Fig. 8.3, the deformation was reduced to 9.07% of the panel thickness, or 0.0299 cm. This value was deemed adequate to support validation of the analytical tool. Since only the center region of the panel was of interest, and after feedback from the industry partners responsible for the development of the analytical tool, the length of the restraint was shortened to include only the central 10.16 cm portion of the panel (5.08 cm above and below the notch).

8.4 Panel Preparation

Test panels with the same geometry and stacking sequence as analyzed were fabricated from unidirectional prepreg tape using the IM7/8552 carbon-epoxy material system [1]. The loaded edges of the panels were bonded to aluminum tabs with EA934 adhesive and several bolts were passed through both the plates and the test article near each end where the panels were mounted in the L-grip fittings. The bolt holes in the tabs can be seen in Fig. 8.1. The outer surfaces of the aluminum tabs were machined to ensure a symmetric cross section.

Both panels were instrumented with electrical resistance strain gages that were oriented axially and placed in back-to-back pairs. The strain gage pattern is shown in Fig. 8.1. Specifically, strain gages were placed in a row along the top of each panel and at the notch tips. The notch tip gages were strip gages where each strip gage consisted of five gages adjacent to each other in one line. These gages were located as close to the notch tip as possible, placing the center of the strain gage nearest the notch tip 1.02 mm from the notch edge. Additional gages were placed outboard mid-length away from the notch. The strain gage locations in the strip gages at the notch tips are shown in Fig. 8.4.

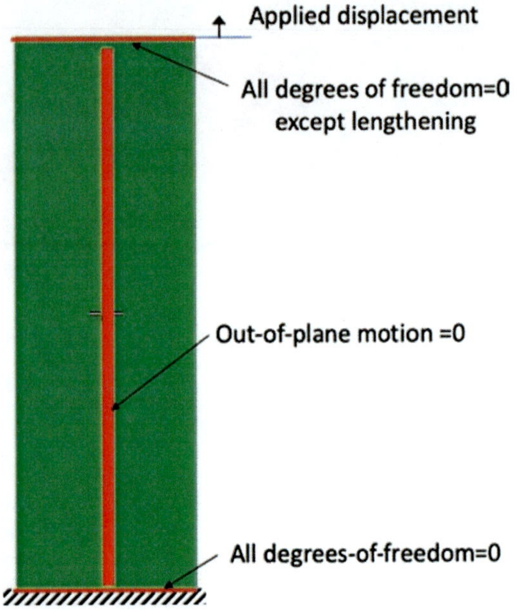

Fig. 8.3 Out-of-plane restraints

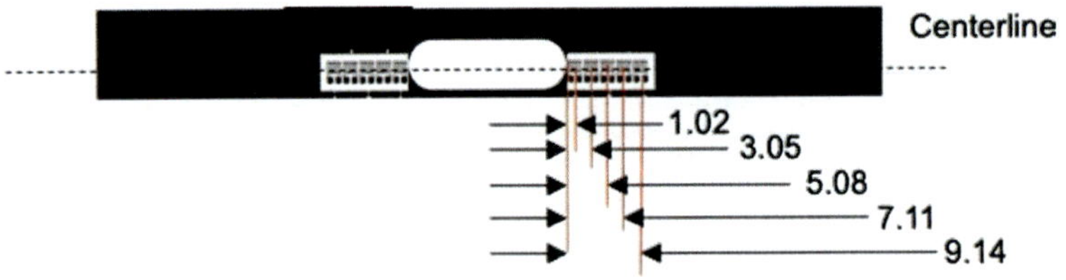

Fig. 8.4 Strain gage locations in strip gages relative to the notch tip with dimensions in mm (not to scale)

8.5 Testing Arrangement and Procedure

Panels were subjected to uniaxial tensile loading via displacement control. A photograph of one panel in the test machine and sketch of a panel in the test configuration are shown in Fig. 8.5. Note that the same fixture arrangement was used on the front and back surfaces of each panel. Load was introduced into the panel through both friction between the aluminum tabs and the L-grip fittings, and through bearing via the bolts at the top and bottom of the panel. The L-grip fittings are shown in gray and the bolts are shown in green in Fig. 8.5b. The aluminum tabs are hidden in Fig. 8.5b by the L-grip fittings, but are shown in Fig. 8.1. Panels were centered in the test machine and loaded to failure through displacement control at a rate of 0.254 mm/min. Changes in panel length were monitored with displacement transducers to capture panel axial displacement independent of test machine mechanism influences. Displacement transducers were placed to the left and right of the panel, in line with the panel, as shown in Fig. 8.5. Note that the "measurement strings" on the transducers are not visible in the photograph in Fig. 8.5a, but they were attached to the bolts at the top of the test fixture on each side of the panel and their locations are highlighted in yellow. The displacement transducers measured from 0.70 m below to 0.81 m above the notch. Out-of-plane deformations were not monitored. The surfaces of central region of the panel were painted white to make cracks more visible for photography.

As discussed earlier, a restraint system to minimize mid-length out-of-plane motion was designed and implemented, and is shown for the front of the panel in the schematic in Fig. 8.5b. In this system, a 10.16-cm-long vertical line at the centerline of the notch on the front and back of the panel was restrained by using two horizontal steel channels (shown in gray) and a short vertical steel bar (shown in red). Only the vertical steel bar at the center of the notch made contact and restrained the panel. The horizontal steel channels supported the vertical bar but were positioned away from the panel surface. Thin Teflon film inserts were placed between the short vertical steel bars at the center of the notch and the panel surfaces to reduce friction with the panel (Fig. 8.6). The longer steel bars parallel to the panel unloaded edges, shown in purple in Fig. 8.5b also did not contact the panel but were instead attached to the top L-grip fittings and were not attached to the bottom L-grip fittings. This arrangement allowed them to move with the top L-grip fitting as the panel lengthened. The thickness of each vertical steel bar was reduced at the bottom end to fit inside the L-grip fittings. Since the inside surfaces of the L-grip fittings are

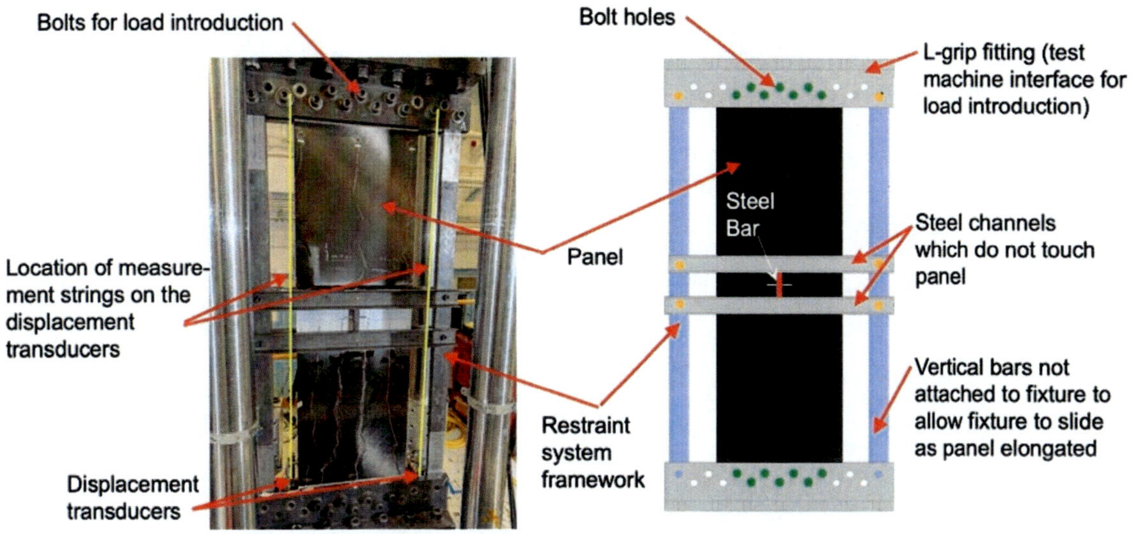

Fig. 8.5 Panel in the test machine. (**a**) Photograph of the panel (**b**) Schematic detailing fixtures

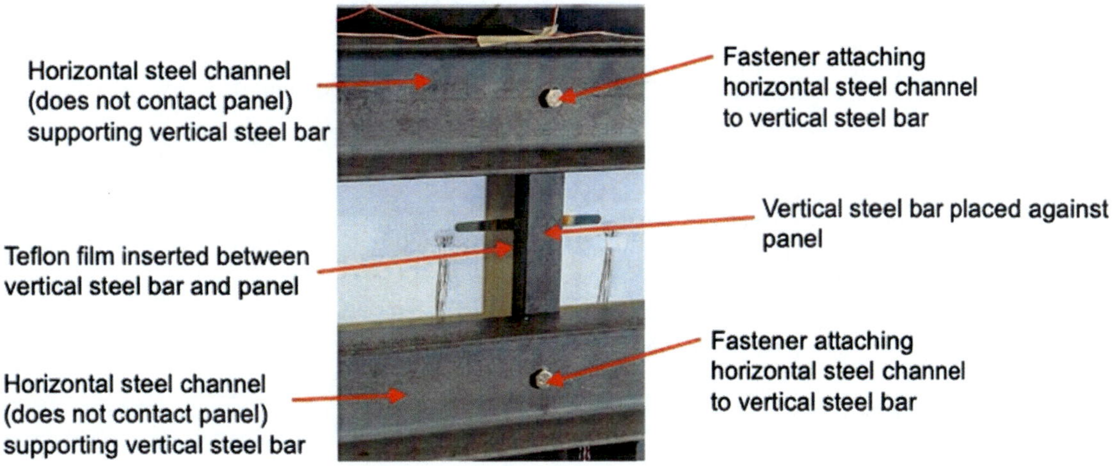

Fig. 8.6 Out-of-plane restraint in the notch area

serrated, Teflon film inserts were placed between the long vertical steel bars and the bottom L-grips fittings to mitigate a potential problem of binding between them to ensure the vertical bars could slide. No binding of these bars was experienced during the tests.

8.6 Results and Discussion

The lengthening of the panel and strains in the vicinity of the notch are presented for nominally identical panels T4-1 and T4-2. Additional strain results are presented in Refs. [1, 2]. This information is used to infer out-of-plane behavior since no direct measurements of out-of-plane deformations were obtained during the tests.

The change in length of each panel, as measured by the displacement transducers on either side of the panel, is presented in Fig. 8.7 where data from panels T4-1 and T4-2 are shown in red and blue lines, respectively. Measurements on the left side of the test panels are shown as the solid lines, and measurements on the right side of the panels are shown as the dashed lines of the same color. All curves show the same slope, and the results are nearly indistinguishable from one another for loads up to 400 kN, indicating that the two panels had the same stiffnesses and the displacements were initially applied uniformly. Results from T4-1 diverge from those of T4-2 at approximately 400 kN, but the left and right sides of T4-1 continue to show the same slope until a load of approximately 415 kN. The load-versus-length results for the left and right of panel T4-2

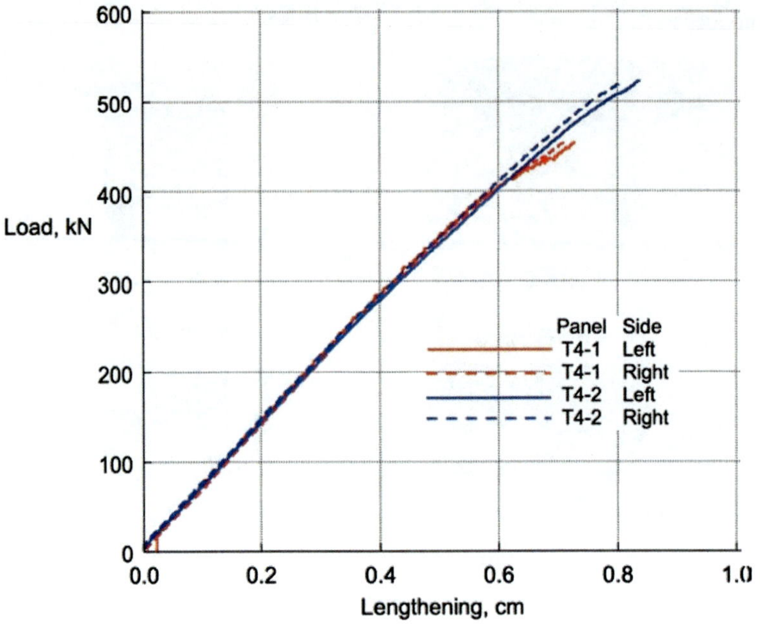

Fig. 8.7 Panel lengthening

remain together until a load of approximately 490 kN. These patterns imply that any failures occurring at loads less than 400 kN were local and did not impact the global response of the panel.

Out-of-plane deformation can be inferred from differences between axial strains on the front and back surfaces of a panel. If there is no difference between the axial strains, there is no bending and no out-of-plane deformation (or only rigid body motion). The greater the difference between the surface strains for the same in-plane location, the more out-of-plane deformation. Only strains measured prior to any local failure can be used in this manner since once local failures have occurred, the laminate may not be intact and/or one or both gages on the surface may be damaged or have separated from the laminate surface. All strains measured 1.02 mm from the notch tip indicate failures of the strain gages or panel surface relatively early in the loading, and therefore, cannot be used to evaluate out-of-plane displacement so they are not included in the following discussion [1, 2].

Strains measured in the gages located 3.05 mm, 5.08 mm, and 7.11 mm from the notch tips within the four strip gages on each panel are shown in Fig. 8.8. Strains at 3.05 mm, 5.08 mm, and 7.11 mm from the notch tips are shown in blue, green, and red, respectively. Strains on the front of the panel are represented by solid lines, while strains on the back of the panel are represented as dashed lines. Each line definition in the figure legend represents two locations, one on the left side and one on the right side of the panel. The consistency among the results for both notch tips of both panels is shown in Fig. 8.8, but in order to evaluate out-of-plane behavior in more depth, it is necessary to break these results apart. Only panel T4-1 will be discussed further since it is representative of both panels.

Strains from these same locations as shown in Fig. 8.8 are shown in Fig. 8.9 for panel T4-1 in a slightly expanded view. In panel T4-1, the load-strain relationship 3.05 mm, 5.08 mm, and 7.11 mm from the notch tips is linear for loads less than approximately 200 kN, 360 kN, and 454 kN (failure), respectively. The difference in measured surface strains at any point in this linear region is within the accuracy of the gage, so no out-of-plane displacement can be detected using these gages. In panel T4-1, at 3.05 mm from the notch tips, some nonlinearity is apparent between loads of 200 kN and 300 kN, but strain continues to increase in all gages and back-to-back strains do not diverge, so no out-of-plane displacement is detected. For greater loads, strains in the front and back surface show disjointed behavior indicative of local failures. In panel T4-1, at 5.08 mm from the notch tips, the gold lines representing the back-to-back pair of gages on the right side of the panel diverge from each other at a load of approximately 360 kN, while the green lines representing the back-to-back pair of gages on the left side of the panel separate at a load of approximately 400 kN. However, both show erratic behavior at these loads which is likely indication of a failure or shift in the structure, so any differences between surface strains cannot be interpreted as representative of out-of-plane displacement. In panel T4-1, at 7.11 mm from the notch tips, all four load-strain relationships show approximately linear results for the entire loading range.

The conclusion to be drawn from these results is that out-of-plane displacement did not occur in either panel. Since neither panel was loaded without the presence of the out-of-plane restraint, the influence of the restraint on the deformation pattern cannot be determined experimentally. Therefore, finite element analysis was used to predict the deformation shape.

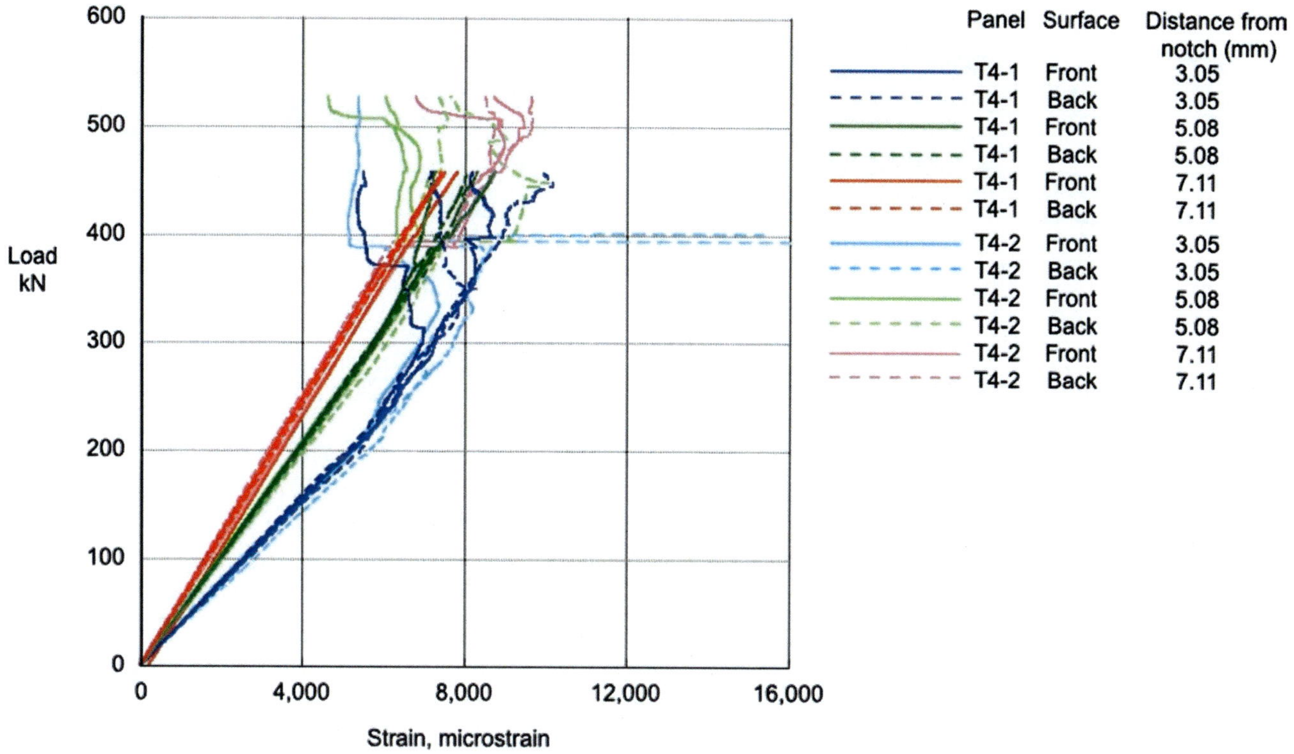

Fig. 8.8 Measured strain 3.05 mm, 5.08 mm, and 7.11 mm from the notch tips in panels T4-1 and T4-2

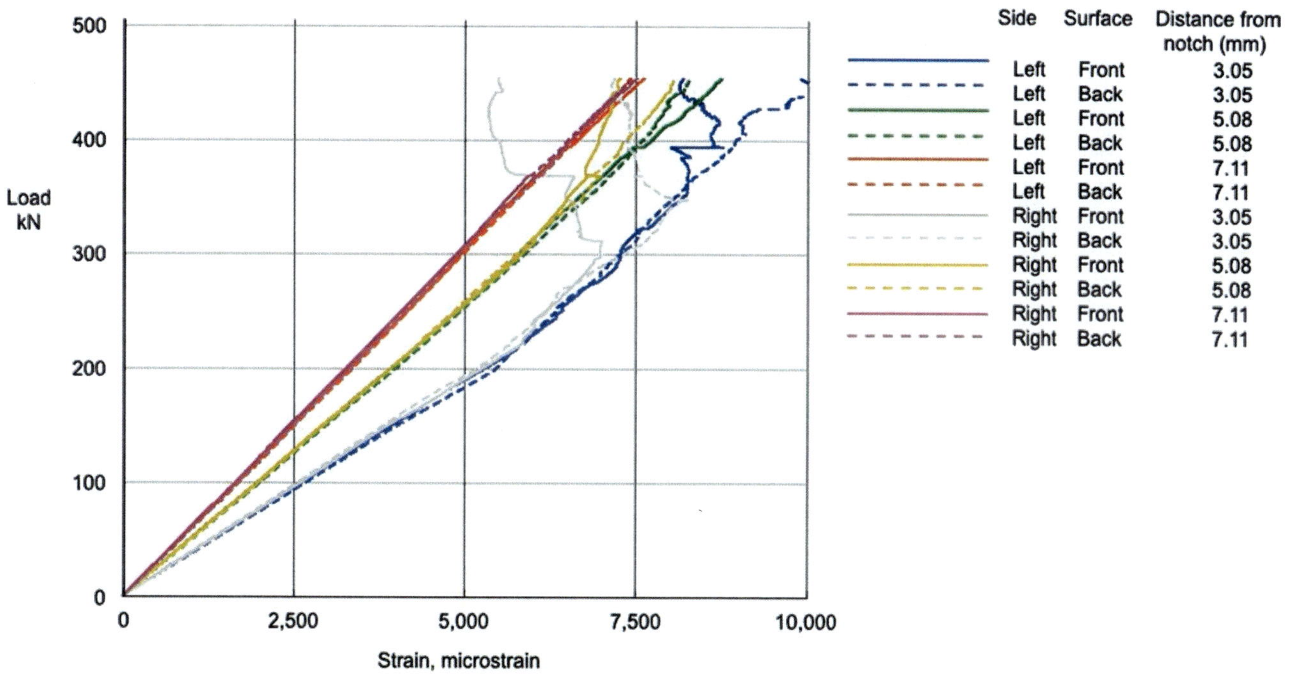

Fig. 8.9 Strain 3.05 mm, 5.08 mm, and 7.11 mm from the notch tips in panel T4-1

Since the initial geometric imperfections were not measured prior to testing, an antisymmetric imperfection pattern was assumed and these analysis results are shown in Fig. 8.10. The change in deformation pattern is shown in Fig. 8.10a, b where the deformation pattern is without the out-of-plane and with the out-of-plane restraint, respectively. These two contour plots are on the same scale so it is clear that the restraint produces very localized changes to the deformation pattern at the notch but has little impact elsewhere in the panel.

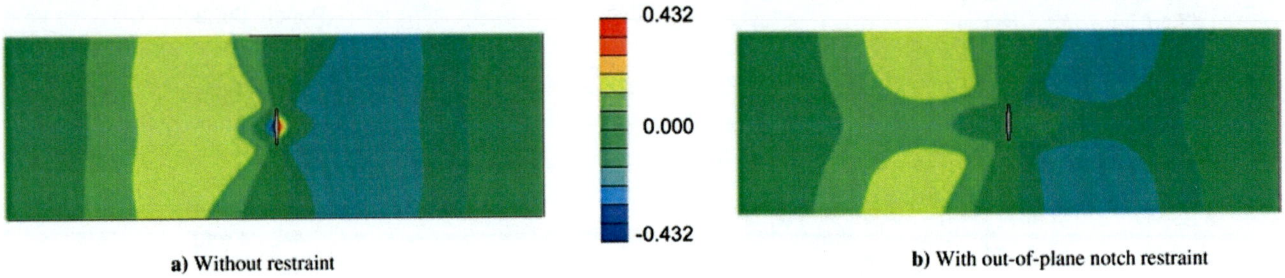

Fig. 8.10 Out-of-plane displacement contours with scale in mm. (**a**) Without restraint (**b**) With out-of-plane notch restraint

8.7 Concluding Remarks

A simple methodology was developed and successfully implemented to suppress out-of-plane deformations, including buckling, for long unstiffened panels with central mid-length notches. This system allowed the panel to slide in-plane within the restraint system without reducing the opportunity to include strain gages within the restrained area and allowed visibility to document the initiation and progression of failure from the central notch.

References

1. Jegley, D.C., Waters Jr., W.A., Zahn, A.: Testing of Composite Laminate Specimens with Mid-Length Lateral Notches. NASA TM-2020-5001774, May 2020
2. Zahn, A., Lovejoy, A.: Test and Analysis Correlation of Standard and Hybrid Standard/Thin-Ply Composite Notched Test Specimens. AIAA SciTech Forum 2020, Orlando, FL, January 6–10, 2020, Paper No. AIAA-2020-0478
3. Enjuto, P., Walker, T.H., Lobo, M., Cregger, E., Wanthal, S.: Investigation of Stiffening and Curvature Effects on the Residual Strength of Composite Stiffened Panels with Large Transverse Notches. AIAA SciTech Forum 2018, Kissimmee, FL, January 8–12, 2018, Paper No. AIAA-2018-2251
4. Abaqus Analysis User's Manual 6.10. Dassault Systèmes, 2010. Accessed on 8 Mar 2019. [Online] Available https://www.sharcnet.ca/Software/Abaqus610/Documentation/docs/v6.10/books/usb/default.htm?startat=pt05ch21s03abm43.html
5. Jones, R.M.: Buckling of Bars, Pates and Shells. Bull Ridge Publishing, Blacksburg (2006)

Chapter 9
Preconditioning Layers Affect Osteoblastic Cell Adhesion to Orthopedic Implant Surfaces

James D. Boyd and Martha E. Grady

Abstract Bacteria can proliferate orthopedic implants, resulting in infection rates as high as 5%. A consistent problem across implantology is the development of surfaces which successfully promote the adhesion and propagation of healthy fibroblast and osteoblast cells while deterring formation of bacterial biofilms. Selecting surface configurations which favor cell adhesion will lead to decreased infection rates. Progress in identifying appropriate surface configurations is hindered by the lack of quantitative adhesion techniques capable of comparing adhesion of cells and biofilms directly. Recent advancements in adhesion techniques have allowed for quantitatively measured adhesion strengths of both bacterial biofilms and cell monolayers using the laser spallation technique. The quantified stress-based adhesion values allow surface and environmental factors that modulate both bacterial and cell adhesion to implant surfaces to be evaluated. During implantation, blood propagates wound sites completely coating implant surfaces. Quantitatively determining the impact of preconditioning layers that accumulate on the implant surface on cell adhesion is vital to predict implant behavior. Previous work has demonstrated that these preconditioning layers either negatively or neutrally impact bacterial adhesion to titanium implant surfaces. This study focuses on the impact that blood plasma and fibronectin coatings have on the adhesion of osteoblastic (MG 63) cells and fibroblasts to the same titanium surfaces. Adhesion results indicate that preconditioning layers and increased surface roughness positively impact cell adhesion. Incorporating the increased adhesion values for cell adhesion into the Adhesion Index demonstrates that increased surface roughness, coupled with natural wound healing preconditioning of surfaces, yields positive biocompatibility.

Keywords Biofilms · Laser spallation · Adhesion · MG 63 · Blood plasma · Surface treatment

Orthopedic implants have infection rates as high as 5% [1–3]. These infections stem from the proliferation of bacteria into bacterial biofilms and in extreme cases can lead to loss of implant [4]. One problem that continually eludes implant developers is creating surfaces which adversely impacts bacterial adhesion while promoting attachment of healthy mammalian cells. Recent advancements in adhesion techniques have allowed for quantitatively measuring the adhesion strength of both bacterial biofilms and cell monolayers using the laser spallation technique [5–7]. The direct comparison now possible allows for the effects of surface and environmental characteristics to be understood and adjusted in order to achieve more favorable adhesion results for cells. However, previous research in this area lacks in vivo that would appropriately represent implant and wound healing environments.

During wound healing, the first thing to contact an implant surface is blood [8, 9]. Since conditioning layers play a significant role in the initial adhesion and proliferation of cells during the wound healing process, the inclusion of relevant plasma proteins is vital to accurately predict cellular growth response onto implantable devices. In this work, the laser spallation technique is employed to measure the adhesion differential between two different bacterial biofilms and osteoblast-like cells on implant-mimicking surfaces with a conditioning layer of either blood plasma or fibronectin. Figure 9.1a illustrates the laser spallation process and detachment of biological film layer while maintaining conditioning layer.

Cell adhesion onto orthopedic implant-mimicking surfaces is the main focus of this study. MG 63 (ATCC CRL-1427), and immature osteoblastic cell line, is cultured onto preconditioned titanium substrates. Titanium is selected as it is one of the most common orthopedic implant materials [10]. Similarly, the substrate surface is roughened to simulate typical orthopedic implant surface roughness [10]. The titanium substrates are then preconditioned with either human blood plasma,

J. D. Boyd (✉) · M. E. Grady
Department of Mechanical Engineering, College of Engineering, University of Kentucky, Lexington, KY, USA
e-mail: jdbo227@g.uky.edu; m.grady@uky.edu

V. Chalivendra, F. Gardea (eds.), *Mechanics of Composite, Hybrid & Multi-functional Materials, Volume 5*, Conference Proceedings of the Society for Experimental Mechanics Series, https://doi.org/10.1007/978-3-031-17445-2_9

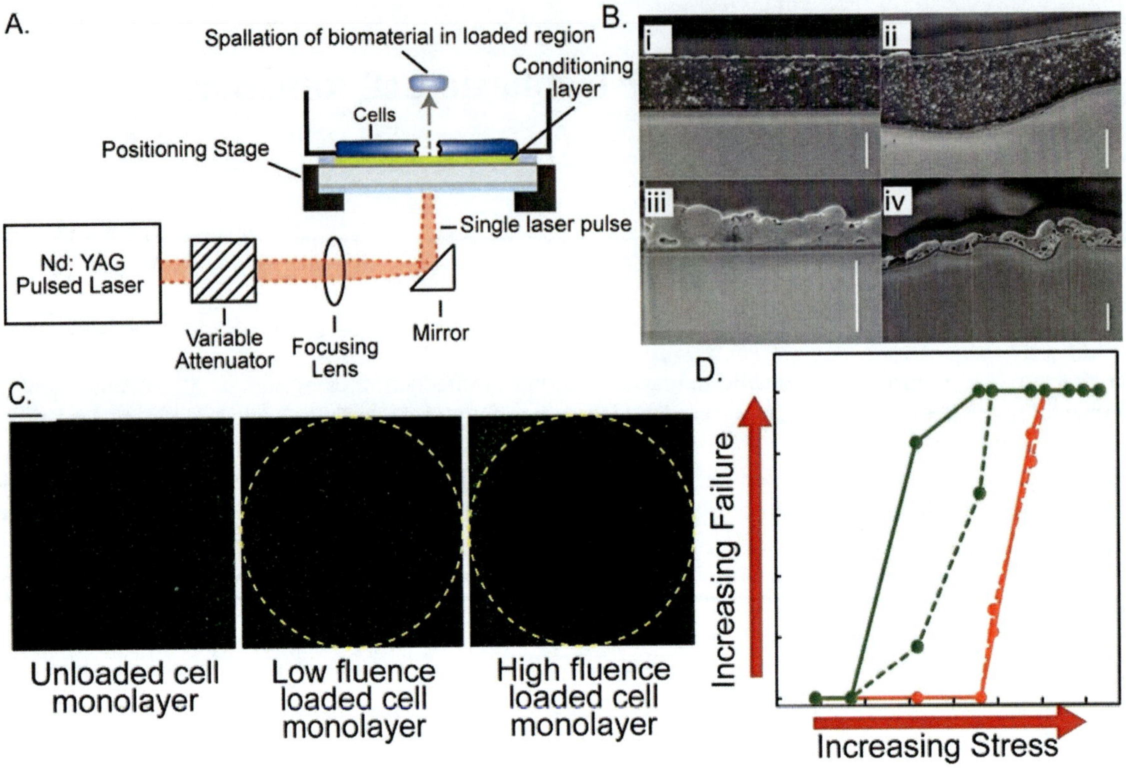

Fig. 9.1 (**a**) The laser spallation set up used during delamination experiments. The Nd:YAG laser used reflected 90° to allow for horizontal orientation of the substrate assembly used, adapted from [12]. (**b**) (i, ii) Depicts SEM images of cross sections for plasma-coated titanium, (iii, iv) are SEM images of cross sections for fibronectin-coated titanium surfaces. Scale bars represent 1 μm vertically. (**c**) Fluorescent microscopy images of MG 63 cells at various loading fluence values. Calcein AM green stain is applied to visualize cells. Yellow circle depicts the 2 mm diameter loaded region. (**d**) Graph of increasing failure of films as a result of increasing interface stress. Green line depicts fibronectin-coated samples while red indicates plasma-coated samples. Solid lines represent smooth titanium surfaces and dashed lines represent roughened surfaces

diluted to 55% by volume, obtained from the University of Kentucky's Biospecimens Core, or 10 μg/ml of human fibronectin, dissolved in PBS. Both preconditioning substances are applied to separate substrates for 1 h before being aspirated to allow cell culturing. Figure 9.1b illustrates representative protein layer thicknesses obtained during the preconditioning process. Scanning electron microscopy (SEM) images were taken from cross-sectioned coating and a consistent layer, less than 2 μm in thickness, was measured. Cells are then cultured on top of preconditioned titanium substrates, using supplemented Eagle's Minimum Essential Medium (EMEM, ATCC 30-2003) for 24 h before spallation experiments occur. Calcein AM (Thermo Fisher Scientific L3224) is applied to cell monolayer to visibly observe attachment and spallation of cells.

The laser spallation technique is a thin film adhesion technique recently applied to biological materials [5, 6, 11]. In our set up, a single laser pulse from an Nd:YAG, wavelength of 1064 nm, impinges upon the backside of our substrate surface. The laser pulse is converted into a mechanical wave through plasma gasification. The compressive wave travels through the substrate-cell systems before reflecting off of the free surface and loading the cell monolayer-titanium interface with a tensile load. Each cell layer system can be loaded multiple times resulting in adhesion quantification of the cell titanium system. The tensile loading of the cells, at sufficient energy, will result in detachment, or spallation, of the cells. The laser energy can be varied in order to find the minimum energy needed to initiate detachment. Figure 9.1c depicts what typical cell regions look like when loaded with both insufficient stress, and sufficient stress to spall from the surface.

As is the standard with laser spallation experiments, a Michelson-type interferometer is used to measure free surface displacement during the loading of substrates. Un-treated titanium substrates are loaded with the same range of fluence, energy per area, and values as used in the spallation experiments to determine the free surface velocity during detachment. Calibration equations are applied to obtain the substrate stress profile. The substrate stress profile is then coupled with transmission and reflection wave coefficients in order to determine the interface stress of the cells and titanium surface. Identical procedure has been applied in Boyd et al. [6].

Four different preconditions were tested in this set of experiments, all cultured with MG 63, including: smooth titanium coated in 55% blood plasma, smooth titanium coated in 10 μg/ml of human fibronectin, rough titanium coated in 55% blood plasma, and rough titanium coated in 10 μg/ml of human fibronectin. Almost 50 separate substrates were culture, with each substrate loaded more than 10 times with increasing fluence values. The failure statistics for these experiments are reported in Fig. 9.1d. Early results indicate a significant difference in adhesion between substrates coated with human plasma and those coated with fibronectin. More stress is required to initiate the detachment of cells on the surface pretreated with human plasma than those with fibronectin. Additionally, there appears to be no correlation with the surface roughness used in this study and adhesion on human plasma-coated substrates. When examining the adhesion of the cells onto fibronectin-coated surfaces, the surface roughness yields an increase in stress needed to initiate detachment. In previous studies, the increase in macroscopic adhesion resulted in a large increase in adhesion strength for MG 63 onto titanium surfaces. The change of the impact of surface roughness on adhesion is believed to be caused by the surface topography change that occurs when preconditioning layers are applied to the surface. Similarly, when compared to the adhesion statistics of uncoated samples it is observed that adhesion decreases when the surface proteins are attached. Further work seeks to compare the adhesion differential these preconditioning layers play with not only mammalian cells but also deleterious oral bacteria.

Acknowledgments This material is based upon work supported by the National Science Foundation CAREER Award grant number 2045853. Any opinions, findings, and conclusions or recommendations expressed in this material are those of the authors and do not necessarily reflect the views of the National Science Foundation.

References

1. Zimmerli, W., Sendi, P.: Orthopaedic biofilm infections. APMIS. **125**(4), 353–364 (2017)
2. Zimmerli, W., Trampuz, A., Ochsner, P.E.: Prosthetic-joint infections. N. Engl. J. Med. **351**(16), 1645–1654 (2004)
3. Phillips, J., et al.: The incidence of deep prosthetic infections in a specialist orthopaedic hospital: a 15-year prospective survey. J. Bone Joint Surg. **88**(7), 943–948 (2006)
4. Arciola, C.R., Campoccia, D., Montanaro, L.: Implant infections: adhesion, biofilm formation and immune evasion. Nat. Rev. Microbiol. **16**(7), 397 (2018)
5. Boyd, J.D., Korotkova, N., Grady, M.E.: Adhesion of biofilms on titanium measured by laser-induced spallation. Exp. Mech. **59**(9), 1275–1284 (2019)
6. Boyd, J.D., et al.: Biofilm and cell adhesion strength on dental implant surfaces via the laser spallation technique. Dent. Mater. **37**(1), 48–59 (2021)
7. Hu, L., et al.: Cell adhesion measurement by laser-induced stress waves. J. Appl. Phys. **100**(8), 084701 (2006)
8. Kirsner, R.S., Eaglstein, W.H.: The wound healing process. Dermatol. Clin. **11**(4), 629–640 (1993)
9. Velnar, T., Bailey, T., Smrkolj, V.: The wound healing process: an overview of the cellular and molecular mechanisms. J. Int. Med. Res. **37**(5), 1528–1542 (2009)
10. Martin, J., et al.: Effect of titanium surface roughness on proliferation, differentiation, and protein synthesis of human osteoblast-like cells (MG63). J. Biomed. Mater. Res. **29**(3), 389–401 (1995)
11. Ehsani, H., et al.: Evolution of the laser-induced spallation technique in film adhesion measurement. Appl. Mech. Rev. **73**(3), 030802 (2021)
12. Boyd, J.: Cell and Biofilm Adhesion Strength Quantified by the Laser Spallation Technique Yields a Novel Biocompatibility Metric, the Adhesion Index, p. 185. Theses and dissertations – mechanical engineering (2021) https://uknowledge.uky.edu/me_etds/185

Chapter 10
Development of a Polyamide 6-Based Composite Material for UAV Propellers

Aybeniz Seyhan, Alaeddin Burak Irez, and Yusuf Polat

Abstract Unmanned aerial vehicles (UAVs) are used in many industries thanks to their beneficial characteristics and propellers are one of the most important components of UAVs. Propellers transform rotary motion into linear thrust and they generally provide lift and thrust to the UAV by spinning and creating an airflow, which results in a pressure difference between the top and bottom surfaces of the propeller. During the life cycle of the propellers, they may receive impact loadings as a result of hard landings or bird strikes. Therefore, impact damage can occur in UAV propellers. Due to that reason, the impact resistance of the propellers must be improved. In this study, a composite material is designed to be used in the manufacture of UAV propellers. Polyamide 6 (PA6) is selected as the matrix material thanks to its superior abrasion resistance, outstanding dimensional stability, and high specific tensile strength. In order to improve the impact resistance of the PA6, olefin block copolymer (OBC) and carbon-based reinforcement including carbon nanotube (CNT) are planned to be used. Following the development of the composite formulation, composites are going to be manufactured and tensile tests, Charpy impact, and hardness tests were carried out to determine fundamental mechanical properties. Finally, optical microscopy was used to examine the CNT distribution in the manufactured composites.

Keywords Polyamide 6 · UAV propeller · Impact resistance · CNT · Thermoplastic composite

10.1 Introduction

The trend for the polymer matrix composites is increasing day by day for engineering applications because of their specific strength and modulus [1]. Thermoplastic polymer-based composites are now gaining more popularity. One of the most promising application fields regarding polymer nanocomposites is various elements of aviation and defense industry. However, they still require further improvement for durability in the high-tech applications in order to replace certain metals [2]. The main drawback of the thermoplastics is their limited mechanical properties.

Authors "Aybeniz Seyhan" and "Alaeddin Burak Irez" have equally contributed to this chapter.

A. Seyhan
Department of Textile Engineering of ITU and TEMAG Lab, Faculty of Textile Technology and Design, Istanbul Technical University, Istanbul, Turkey

TEMAG Labs, Textile Technology and Design Faculty, Istanbul Technical University, Istanbul, Turkey
e-mail: seyhanay@itu.edu.tr

A. B. Irez (✉)
Department of Mechanical Engineering of ITU, Faculty of Mechanical Engineering, Istanbul Technical University, Istanbul, Turkey
e-mail: irez@itu.edu.tr; onsel21@itu.edu.tr

Y. Polat
TEMAG lab and Erzurum Technical University, Textile Technology and Design Faculty, Istanbul Technical University, Istanbul, Turkey

Department of Mechanical Engineering, Faculty of Engineering and Architecture, Erzurum Technical University, Erzurum, Turkey
e-mail: yusuf.polat@erzurum.edu.tr

V. Chalivendra, F. Gardea (eds.), *Mechanics of Composite, Hybrid & Multi-functional Materials, Volume 5*, Conference Proceedings of the Society for Experimental Mechanics Series, https://doi.org/10.1007/978-3-031-17445-2_10

Polyamide is one of the most commonly used polymers which has 70 years of usage in different applications. Among other types, researchers paid attention to polyamide 6 (PA6) because of its high strength, good fatigue performance, and high melting point [3]. However, PA6 exhibits limited toughness when exposed to impact loading due to its high glass transition temperature. To improve the impact behavior of PA6, a secondary mobile phase such as elastomeric compounds needs to be incorporated within the polymer [4, 5]. The effect of polymer blends [6–9] and fiber reinforcements such as short carbon fiber and glass fiber [10] on the mechanical properties of the composite structures have been studied in order to enhance the mechanical performance of polymer matrix composites. Today, nanomaterials have presented new functionalities to novel composites for various application areas [11]. Carbon nanotubes (CNTs) found place in many engineering areas due to their outstanding properties like strength, elastic modulus, thermal and electrical conductivity, and high aspect ratio [12]. The CNT reinforcement makes the strength, modulus, and microhardness of PA6 higher [12].

In this study, multi-walled carbon nanotube (MWCNT) and olefin block copolymer (OBC) as an elastomeric polymer were incorporated to polyamide 6 to enhance the tensile properties and energy absorption properties of composites. In order to overcome the phase separation in the polymer matrix, impact modifier OBC was grafted with maleic anhydride. The effect of CNT content on OBC/PA6 matrix system has been investigated.

10.2 Materials and Methods

The anamide 33B grade polyamide 6 (Anadolu Iplik, Turkey) was used as thermoplastic resin. As the first modifier, olefin block copolymer, INFUSE 9100 was supplied from DOW Chemical Company. The density, the melting temperature, and MFI of OBC are stated as 0.877 g/cm^3, 130 °C, and 1 g/10 min, respectively. The carboxylic acid functionalized MWCNT with >96% purity and 28–48 nm outer diameter was purchased from Nanokar company (Turkey). Granular maleic anhydride (Merck, Germany), granular dicumyl peroxide (Sigma Aldrich, Germany), and analytic grade acetone (Isolab, Turkey) were supplied for grafting procedure.

Composites were manufactured with compositions given in Table 10.1. The schematic shown in Fig. 10.1 was used for the manufacturing. Maleic anhydride (MA) and dicumyl peroxide (DCP) were dissolved, CNT was dispersed in acetone.

Table 10.1 Compositions of composite samples

	wt. % of OBC	wt. % of CNT
PA6	0	0
PA6 + OBC	10	0
PA6 + OBC + MAH	10	0
PA6 + OBC + CNT3	10	0.3
PA6 + OBC + CNT4	10	0.4
PA6 + OBC + CNT5	10	0.5

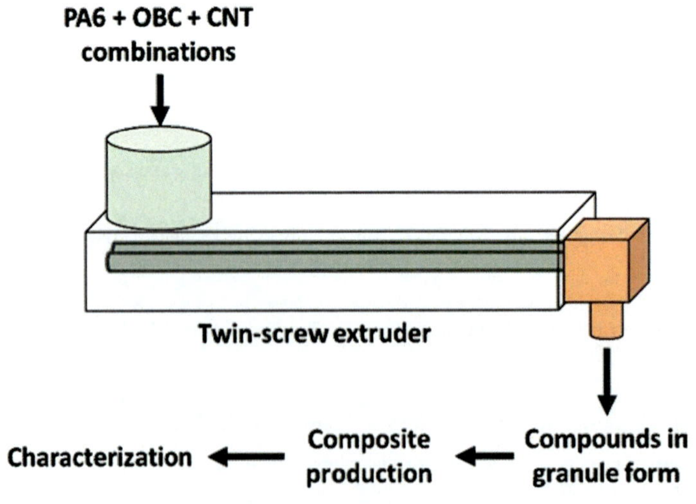

Fig. 10.1 Schematic illustration of composite sample production steps

OBC was added to the mixture. Then solvent was evaporated at 50 °C. The OBC and CNT were mixed to PA6 by using a twin-screw extruder. After the extrusion process, the compounds were cut into granule form. The compound granules were fed to a plastic injection system to obtain composite samples.

10.2.1 Material Characterization

The test samples were prepared after composite production. The tensile test, Charpy impact test, and hardness tests were realized. ASTM D638 standard was used for the tensile test. At least five specimens were prepared from each group of samples. Tensile tests were carried out using a Shimadzu AG-x testing machine with a loading rate of 10 mm/min using a 50 kN load cell. ASTM 6110 standard was applied to Charpy impact samples by using a universal Charpy tester of Zwick Roell HIT5P to see the impact resistance of samples. The dimensions of impact test samples are 125 mm × 12.7 mm rectangular shape with a standard notch. For optical microscopy, the sections of composites were cut and they were cold mounted. Polishing was performed. Optical microscopy was conducted to observe the microstructure of the manufactured composites and CNT distribution was observed.

10.3 Results and Discussion

The microscopical observation was conducted to analyze the CNT distribution within the PA6 matrix. The black-colored regions surrounded with red circles in Fig. 10.2 show the CNT agglomerations. Other areas show the homogeneously distributed regions of CNTs in resin.

After microscopic observation, densities of each group were measured and they are given in Table 10.2. According to data in Table 10.2, density values decreased with OBC and CNT addition.

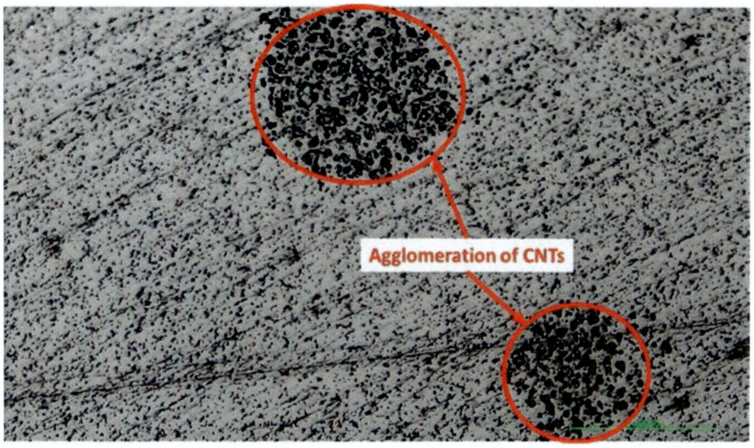

Fig. 10.2 Microstructural observation of CNT-added samples and agglomeration

Table 10.2 Densities of the produced composites

Composition	Density (g/cm^3)
PA6	1.1287 ± 0.009
PA6 + OBC	1.1061 ± 0.008
PA6 + OBC + MAH	1.103 ± 0.001
CNT3	1.1129 ± 0.007
CNT4	1.1106 ± 0.009
CNT5	1.1157 ± 0.009

Table 10.3 Tensile test results of composite samples

Specimen group	Elasticity modulus (MPa)	UTS (MPa)	Strain at break
PA6	3005.89 ± 107.95	76.85 ± 2.20	0.756 ± 0.145
PA6 + OBC	2704.06 ± 75.15	69.47 ± 3.65	0.728 ± 0.129
PA6 + OBC + MAH	2838.46 ± 174.65	68.96 ± 2.63	1.033 ± 0.2
PA6 + OBC + CNT3	3403.54 ± 155.72	74.97 ± 1.24	0.292 ± 0.92
PA6 + OBC + CNT4	3110.63 ± 242.77	73.26 ± 2.95	0.256 ± 0.087
PA6 + OBC + CNT5	2914.32 ± 185.80	71.10 ± 5.43	0.184 ± 0.093

Table 10.4 Charpy impact test results (J/mm^2)

PA6	PA6 + OBC	PA6 + OBC + MAH	PA6 + OBC + CNT3	PA6 + OBC + CNT4	PA6 + OBC + CNT5
0.59 ± 0.23	0.98 ± 0.14	1.33 ± 0.15	0.9 ± 0.18	0.77 ± 0.35	0.78 ± 0.07

Table 10.5 Hardness test results (Shore D)

PA6	PA6 + OBC	PA6 + OBC + MAH	PA6 + OBC + CNT3	PA6 + OBC + CNT4	PA6 + OBC + CNT5
75.33 ± 0.51	70.5 ± 3.1	63.83 ± 4.71	73.67 ± 1.63	74.33 ± 0.82	75 ± 0.00

After density measurement, mechanical characterizations were performed. The tensile test results are presented in Table 10.3. The OBC addition decreased the tensile properties by nearly 10% for both modulus and ultimate tensile strength (UTS). Maleic anhydride grafting of OBC improved the modulus which means stiffness of material increased while UTS remained almost the same. However, strain at break also increased by grafting. In addition to OBC, CNT was more effective to enhance the strength and modulus of composite samples. The addition of 0.3 wt.% CNT increased the modulus 11.7% while UTS did not change significantly. The modulus improvement resulted in a decrease in the strain at break. In this respect, when the CNT percentage increased, the mechanical properties were possibly decreased because of the agglomeration of CNTs as shown in Fig. 10.2. Among all samples, the PA6 + OBC + CNT3 sample is the best for the tensile properties.

Charpy impact tests were performed for all samples. The results are given in Table 10.4. The results showed the 10 wt.% OBC addition increased the energy absorption by nearly 66%. Furthermore, MA grafting has a complimentary effect on OBC which provided an increase of 125%. The interface between OBC and PA6 was possibly improved by grafting which resulted in more load transfer from polyamide to elastomeric polymer OBC. However, CNT addition which is increasing the stiffness of samples decreased the impact energy when compared with maleic grafted samples. Still the energy absorption value of 0.3 wt.% CNT incorporated sample is 52% higher than pure PA6. The more the CNT concentration the lower the absorbed impact energy.

Finally, the shore D hardness test was conducted for all samples. The OBC addition decreased the stiffness of samples, similarly the hardness values were also decreased. On the contrary, CNT addition increased the stiffness together with hardness (Table 10.5).

10.4 Conclusion

As a conclusion, thermoplastic composites were developed for the application of unmanned aerial vehicle propellers. The thermoplastic composite materials were produced by using a twin-screw extruder. Then, the injection molding technique was used for composite sample production. The microstructural observation showed the CNT distribution and agglomeration. Before mechanical tests, the density of each group was measured. The CNT and OBC addition decreased the density slightly. Tensile, Charpy impact, and hardness tests were performed. The results of these tests showed that the OBC and CNT addition to polyamide 6 is suitable for application of the UAVs propeller. The results obtained in lab-scale are promising and by using more sophisticated laboratory facilities, the problems in the homogeneity can be eliminated and the results can be more improved.

Acknowledgments The authors thank Dow Chemicals Company and Anadolu Iplik for their material support and valuable discussions in certain stages of this project. In addition, the authors acknowledge the support of Alparslan Erdirencelebi, Gunay Maltas, and Serdar Babur during the composite manufacturing.

References

1. Pisani, W.A., Wedgeworth, D.N., Roth, M.R., Newman, J.K., Shukla, M.K.: Computational prediction of mechanical properties of PA6–graphene/carbon nanotube nanocomposites. J. Phys. Chem. C. **125**, 15569–15578 (2021). https://doi.org/10.1021/acs.jpcc.1c03410
2. Petrény, R., Mészáros, L.: The effect of microstructure on the dynamic mechanical properties of carbon fiber and carbon nanotube reinforced multi-scale composites with a polyamide 6 matrix. IOP Conf. Ser.: Mater. Sci. Eng. **903**, 012021 (2020). https://doi.org/10.1088/1757-899X/903/1/012021
3. Zaldua, N., Maiz, J., de la Calle, A., García-Arrieta, S., Elizetxea, C., Harismendy, I., et al.: Nucleation and crystallization of PA6 composites prepared by T-RTM: effects of carbon and glass fiber loading. Polymers. **11**, 1680 (2019). https://doi.org/10.3390/polym11101680
4. Huang, J., Schmauder, S., Weber, U., Geier, S.: Micromechanical modelling of the elastoplastic behaviour of nanodispersed elastomer particle-modified PA6. Comput. Mater. Sci. **52**, 107–111 (2012). https://doi.org/10.1016/j.commatsci.2011.01.005
5. Lin, X., Liu, Y., Chen, X., Wu, Y., Cui, L., Mao, L., et al.: Reactive compatibilization of polyamide 6/olefin block copolymer blends: phase morphology, rheological behavior, thermal behavior, and mechanical properties. Materials. **13**, 1146 (2020). https://doi.org/10.3390/ma13051146
6. Huang, J.J., Keskkula, H., Paul, D.R.: Elastomer particle morphology in ternary blends of maleated and non-maleated ethylene-based elastomers with polyamides: role of elastomer phase miscibility. Polymer. **47**, 624–638 (2006). https://doi.org/10.1016/j.polymer.2005.11.087
7. Li, Y., Shimizu, H.: Conductive PVDF/PA6/CNTs nanocomposites fabricated by dual formation of cocontinuous and nanodispersion structures. Macromolecules. **41**, 5339–5344 (2008). https://doi.org/10.1021/ma8006834
8. Ohlsson, B., Hassander, H., Törnell, B.: Improved compatibility between polyamide and polypropylene by the use of maleic anhydride grafted SEBS. Polymer. **39**, 6705–6714 (1998). https://doi.org/10.1016/S0032-3861(97)10290-7
9. Yoo, Y., Tiwari, R.R., Yoo, Y.-T., Paul, D.R.: Effect of organoclay structure and mixing protocol on the toughening of amorphous polyamide/elastomer blends. Polymer. **51**, 4907–4915 (2010). https://doi.org/10.1016/j.polymer.2010.08.036
10. Tjong, S.C., Xu, S.A., Mai, Y.W.: Impact fracture toughness of short glass fiber-reinforced polyamide 6,6 hybrid composites containing elastomer particles using essential work of fracture concept. Mater. Sci. Eng. A. **347**, 338–345 (2003). https://doi.org/10.1016/S0921-5093(02)00609-3
11. Doagou-Rad, S., Islam, A., Jensen, J.S., Alnasser, A.: Interaction of nanofillers in injection-molded graphene/carbon nanotube reinforced PA66 hybrid nanocomposites. J. Polym. Eng. **38**, 971–981 (2018). https://doi.org/10.1515/polyeng-2018-0060
12. Dinesh Kumar, S., Ravichandran, M., Alagarsamy, S.V., Chanakyan, C., Meignanamoorthy, M., Sakthivelu, S.: Processing and properties of carbon nanotube reinforced composites: a review. Mater. Today: Proc. **27**, 1152–1156 (2020). https://doi.org/10.1016/j.matpr.2020.02.006

Check for updates

Chapter 11
In Situ Measurement of AC Conductivity to Quantify Unidirectional Alignment of Graphene Nanoplatelets (GNPs) in Epoxy

Sai Tharun Kotikalapudi and Raman P. Singh

Abstract Graphene nanoplatelets (GNPs) are widely used as conductive fillers for various polymer nanocomposites. However, the actual performance of these materials is much lower than theoretical predictions due to the natural tendency of GNPs to agglomerate. Contemporary techniques for random dispersion do not offer modulation of the orientation of graphene and thereby limit its capabilities, promoting the usage of higher concentrations that can lead to undesirable effects. Electric field alignment is a promising way to obtain significantly improved directional properties with low concentrations of GNPs. This paper presents the use of real-time AC conductivity measurements to characterize the unidirectional alignment process of GNPs in a thermoset epoxy polymer. Theoretical modeling of the effects of alternating electric field on a single transversely isotropic GNP reveals the key parameters that influence the alignment process, namely, the size of the platelet, viscosity of the epoxy, and the content of GNPs. Considering these parameters, an experimental setup is devised to create an alternating electric field in an aluminum mold separated by epoxy/GNP mixture. Time required for alignment is calculated based on the chain formation and rotation time obtained from the above parameters. Preliminary research shows a two- to threefold increase in electrical conductivity in aligned direction compared to randomly dispersed epoxy/GNP samples. The effects of alignment also reveal an increase in the mechanical strength, resulting in a (10–15%) increase in Young's modulus in the aligned direction. These significant property improvements also decrease the percolation threshold of randomly dispersed epoxy/GNP sample from 1 to 0.5 wt%. Dielectric spectroscopy of randomly oriented and aligned samples shows a 100% increase in dielectric constant at a frequency of 1000 Hz in the aligned direction.

Keywords Graphene · Alignment · Conductivity · Dielectrics · Nanocomposite

11.1 Introduction

In recent years, polymer nanocomposites have been continually used in industry due to their ease of processability, low cost, and their propensity to increase electrical and mechanical properties simultaneously [1–3]. Nanoparticles such as graphene and graphene oxide have a unique chemical structure that contributes to its compelling mechanical, electrical, and thermal performance of nanocomposites [4, 5]. Applications for these materials include dielectrics, lightning strike protection systems, supercapacitors, multifunctional composites, and flexible electronics [5–7]. These property increments are achieved by the two-dimensional hexagonal structure of graphene, which has a thickness of one-atom-layer [8]. Graphene nanoplatelets consist of multiple layers of graphene stacked together to form an oblate spheroidal structure. The above-mentioned properties of GNP vary significantly between the perpendicular and parallel directions due to its anisotropic nature [9]. To fully utilize these properties, GNPs orientation must be controlled during the manufacturing process of nanocomposites.

S. T. Kotikalapudi (✉)
School of Mechanical and Aerospace Engineering, Oklahoma State University, Stillwater, OK, USA
e-mail: skotika@okstate.edu

R. P. Singh
School of Mechanical and Aerospace Engineering, Oklahoma State University, Stillwater, OK, USA

School of Materials Science and Engineering, Oklahoma State University-Tulsa, Tulsa, OK, USA
e-mail: raman.singh@okstate.edu

V. Chalivendra, F. Gardea (eds.), *Mechanics of Composite, Hybrid & Multi-functional Materials, Volume 5*, Conference Proceedings of the Society for Experimental Mechanics Series, https://doi.org/10.1007/978-3-031-17445-2_11

One promising way to orient nanoparticles is by using an external electrical field during the curing process of epoxy [10]. Both DC and AC can be used but, DC can lead to electrophoretic deposition near the electrodes of the applied field due to the directed movement of particles, resulting in unwanted deposition [11]. Wu et al. aligned multilayer graphene flakes with an external electric field to improve multifunctional properties of graphene composites such as directional electrical conductivity, thermal conductivity, and fracture toughness [12]. Kwon et al. studied the growth of mechanical and electrical properties under alternating electric field relating the increase in properties to orientation of GNPs [13]. In both cases, theoretical modeling of alternating electric field effects on a single GNP in epoxy resulted in critical parameters for alignment such as time required for alignment. The model was verified by considering optical microscopy images and measuring the direct current flowing through the sample which resulted in frequent interruption of the experiment. However, these models did not account to the viscosity increase associated with the addition of GNPs to the polymer, which significantly affects the time required for alignment. In addition to that, the experiments were performed for very low weight fractions of graphene (0.1–0.64 wt%) and the theoretical predictions fail to apply when higher weight fractions (beyond percolation threshold) are considered.

Here, we present an in-situ technique of measuring the AC current during the alignment process as a way of correlating the alignment time with various GNP concentrations in epoxy. We also emphasize various property increments associated with aligned graphene nanocomposites such as Young's modulus, percolation threshold, electrical conductivity, and dielectric constant.

11.2 Experimental

11.2.1 Materials

The GNPs (M25, XG Sciences, Lansing, Michigan) have an average thickness of 6–8 nm with a surface area of 120–150 m^2/g and an average particle diameter of 25 μm. Epoxy resin (EPON 862, Hexion, Columbus, Ohio) and hardener (EPIKURE Curing agent 3274, Hexion, Columbus, Ohio) were used for dispersing the GNPs. EPON 862 is a low viscosity blend of diglycidyl ether of bisphenol F. The liquid hardener is a moderately reactive low viscosity aliphatic amine.

11.2.2 Sample Preparation

Nanocomposites were made with different weight fractions of GNPs (0.7, 1.4, and 2.8 wt%). First, GNPs in powder form were added to epoxy resin and stirred with a magnetic stirrer for an hour to dissolve and break the larger agglomerates. The mixture was then passed through a three-roll mill (T65, Torrey Hills Technologies, San Diego, California) for 15 times at 70 rpm with a roller-gap distance of 25 μm to obtain uniform dispersion of GNPs in epoxy. After the dispersion, stoichiometric ratio of hardener was added to the mixture and degassed under vacuum conditions for 30 min to remove the trapped air bubbles. The mixture was then poured into a mold made of aluminum plates of size 200 × 200 mm and separated by 3 mm insulating polymer spacers. The two plates are connected to a wideband amplifier (7602 M, Krohnite, Brockton, Massachusetts) which is connected to an AC signal generator (3312A, HP, Alto California). This was connected in series to a multimeter (5491B, BK Precision, Yorba Linda, California). LabVIEW software was used to track the minute changes in AC current. The schematic is represented in Fig. 11.1. AC electric field (35 V/mm, 10 kHz) was applied for a set amount of time depending on the weight fraction of GNP in epoxy and left to cure at room temperature for 24 h.

11.2.3 Characterization

The dynamic viscosity of the GNP and epoxy mixture was measured prior to the alignment process using a viscometer (LVDVE, Brookfield, Middleboro, Massachusetts) according to ASTM-D2196. Post alignment and curing, 10 samples of size 10 × 10 × 3 mm were cut for each weight fraction using a precision saw and were lightly sanded on top and bottom surfaces using 800 grit paper. Due to the presence of insulating epoxy, the top and bottom surfaces were coated with an electrically conductive epoxy (Duralco 120, Cotronics, Brooklyn, New York) to provide uniform electrical contact. The samples were cured at room temperature for 24 h prior to electrical testing. The DC electrical conductivity was measured for

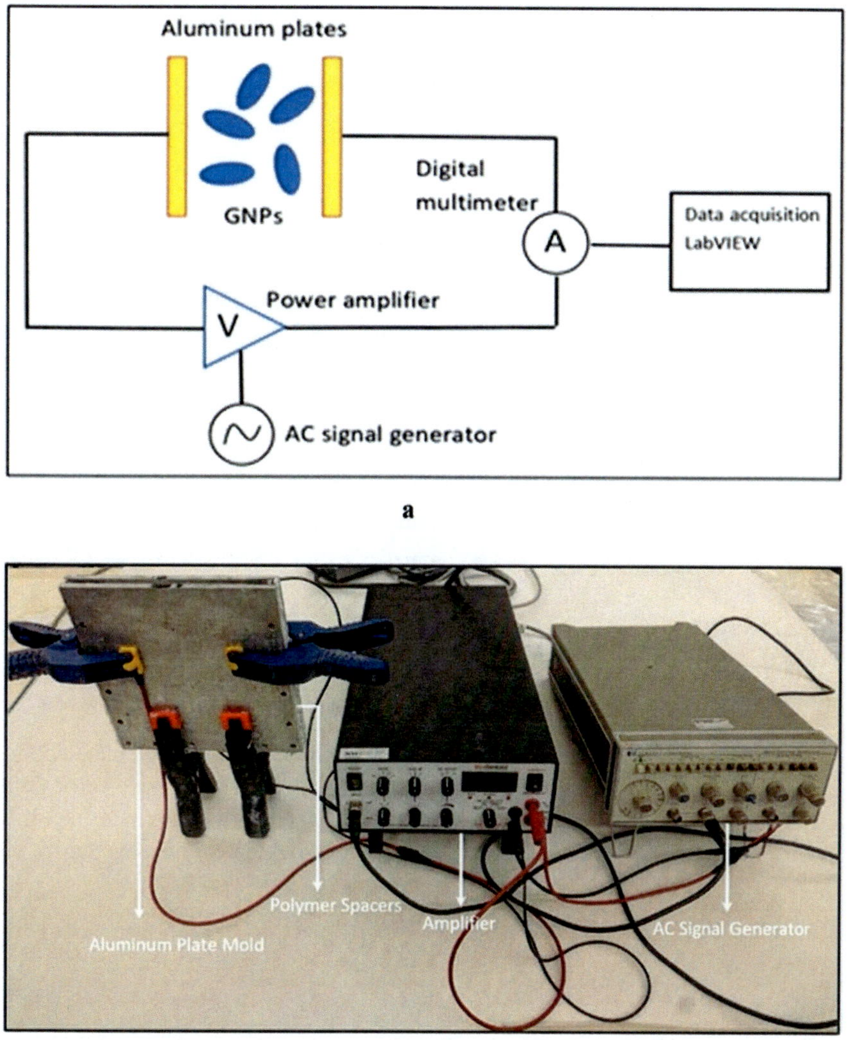

a

b

Fig. 11.1 (**a**) Schematic and (**b**) real experiment setup depicting the alignment process

five samples using a DC power supply connected to a precision multimeter (5491B, BK Precision, Yorba Linda, California). The Young's modulus of the samples in the aligned direction was measured using an ultrasonic pulse-echo method. A negative square wave pulse-receiver (5077PR, Panametrics, Waltham, Massachusetts) was connected to an oscilloscope (3052, Tektronix, Beaverton, Oregon) to measure the ultrasonic sound wave propagation in the aligned direction of GNP. The Young's modulus along the thickness direction was calculated using density, shear, and longitudinal wave speed. In addition to this, dielectric spectroscopy was performed to obtain dielectric constants and AC electrical conductivity along the thickness direction using a potentiostat (VersaSTAT 3F, Ametek, Berwyn, Pennsylvania). For these properties, the frequency was ranged from 10^3 to 10^6 Hz. In all characterization tests, randomly oriented samples were compared to aligned samples. Dielectric constant and dielectric loss were calculated using standard formulations as follows [12]:

$$\left(\text{Dielectric constant}\right)\varepsilon_r^{'} = \frac{-Z''}{2\pi f \varepsilon_0 \left((Z')^2 + (Z'')^2 \right)} \frac{l}{A} \tag{11.1}$$

$$\left(\text{Dielectric loss}\right)\varepsilon_r^{''} = \frac{Z'}{2\pi f \varepsilon_0 \left((Z')^2 + (Z'')^2 \right)} \frac{l}{A} \tag{11.2}$$

Here, f is the AC frequency, ε_0 is the permittivity of free space, l and A are sample thickness and cross-section respectively, and Z' and Z'' are the real and imaginary parts of impedance.

11.3 Results and Discussion

A single GNP within an uncured epoxy experiencing an AC electric field will experience electronic polarization and results in the formation of a dipole. These dipoles are formed due to the difference in dielectric properties of the GNP and the epoxy. Generally, these dipoles are not aligned with the electric field. Due to the oblate spheroidal shape of the GNP, the polarization forces along the platelet are significantly higher than perpendicular to the platelet. The resultant dipole produces a torque about the center of the platelet and tends to rotate toward the electric field. This rotational torque of the particle is opposed by the viscous torque generated by the liquid epoxy. The resultant torque is a function of the electric field, platelet dimensions, viscosity, and dielectric constant of the epoxy. The approximate time required (t) to rotate from an initial angle θ_0 to a final angle θ' was formulated by Wu and coworkers and is given below [12].

$$t = \frac{1}{A}\ln\frac{\tan\theta_0}{\tan\theta'} \tag{11.3}$$

$$A = \frac{\pi}{8\eta}\frac{\varepsilon_m}{\left(\frac{\pi}{2}-\frac{b}{a}\right)}E_0^2 \tag{11.4}$$

Here, A is a function of the viscosity of the epoxy η, dielectric constant of the epoxy ε_m, platelet dimensions a (diameter), b (thickness), and electric field strength E_0. Typically, after the platelet is aligned along the electric field, the particles nearby to each other get attracted due to dipole-dipole attraction and thus forming a continuous chain-like structure from one end of the electrodes to the other. The time required for two platelets to form a connection is given by the following equation devised by Wu et al. [12]:

$$t_C = \frac{2x_o^3}{3B} \tag{11.5}$$

$$x_0 = \frac{1}{\rho}\frac{m_G}{w_G}\frac{1}{4a^2} \tag{11.6}$$

where x_0 is the initial separation distance between two particles considering uniform distribution. Also, m_G is the mass of GNP which is calculated to be around 2.65×10^{-14} kg for a M25 GNP particle and w_G is the weight fraction of GNP in the epoxy. The radius of the platelet is a which is 12.5 μm. The density of the epoxy system was calculated from the densities of the epoxy and hardener used which is 1.1 g/cm^3. Here, B is a constant depending on the viscosity, dimensions of the platelet, dielectric constant of the epoxy, and the intensity of the electric field. The approximate time required for rotation and translation for various weight fractions of GNPs is given in Table 11.1, along with the experimental alignment time obtained from AC conductivity measurements.

Viscosity of the epoxy/GNP system plays a significant role in determining the alignment time. Although an increase in viscosity increases the time required for chain formation, the proximity of particles in higher weight fractions reduces the time for translation. Figure 11.2 shows the AC conductivity variation during the alignment process for various GNP concentrations in epoxy. The experiments are performed twice for each weight fraction to show the repeatability of the AC conductivity results. Initially, before the start of alignment, AC conductivity depends on the amount of GNP in epoxy. Higher GNP content gives rise to higher conductivity values. As alignment is taking place, the AC conductivity increases due to the rotation of the platelets which decreases the resistance. In addition to this, the chain formation of the particles takes place simultaneously during the rotation process. The formation of these conductive chains increases the AC current flowing through the system. The difference from the AC conductivity curves obtained for different weight fractions can be related to the amount the GNPs and the increment in the number of chains as GNP content increases. It can also be noticed that the AC conductivity curves plateau after a certain time denoting the maximum conductivity gain possible in each GNP weight fraction. However, there seems to be a significant mismatch from theoretical predictions and experimental data observed. This can be attributed

Table 11.1 Viscosity of the epoxy/GNP mixture measured by viscometer and subsequent calculations of theoretical rotation time and translation time of particles and experimental alignment time of various GNP weight fractions

GNP wt%	Viscosity (mPas)	Rotation time (min) (Analytical)	Translation time (s) (Analytical)	Alignment time (min) (Experimental)
0.7	1040	22	0.77	50–55
1.4	1400	30	0.13	60–65
2.8	1900	42	0.02	70–80

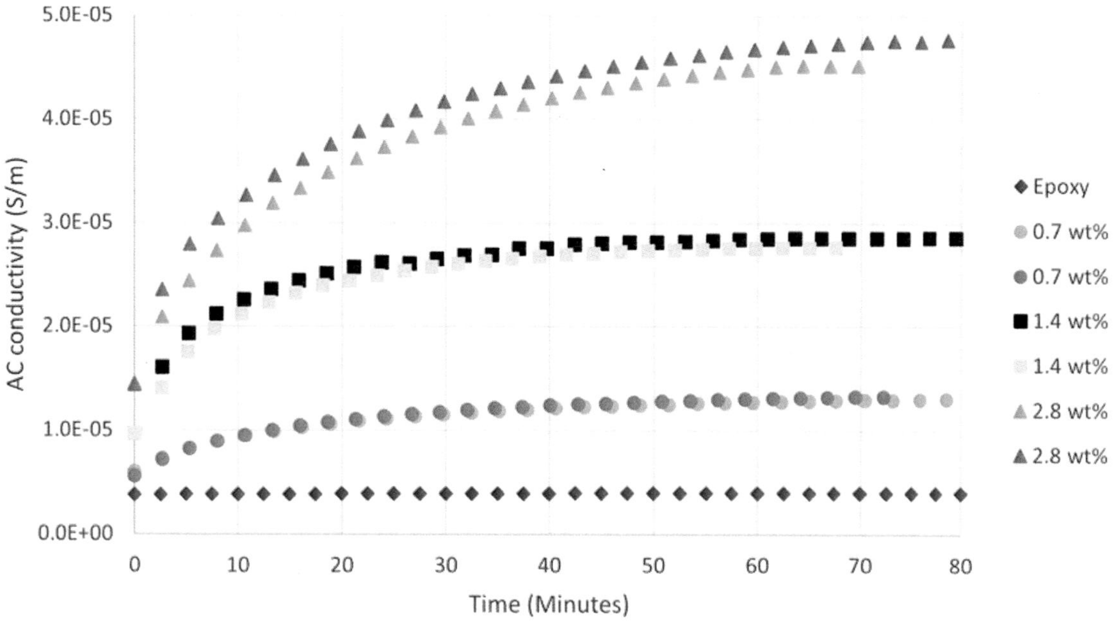

Fig. 11.2 Variation of AC conductivity for different weight fractions of GNPs in epoxy with alignment time

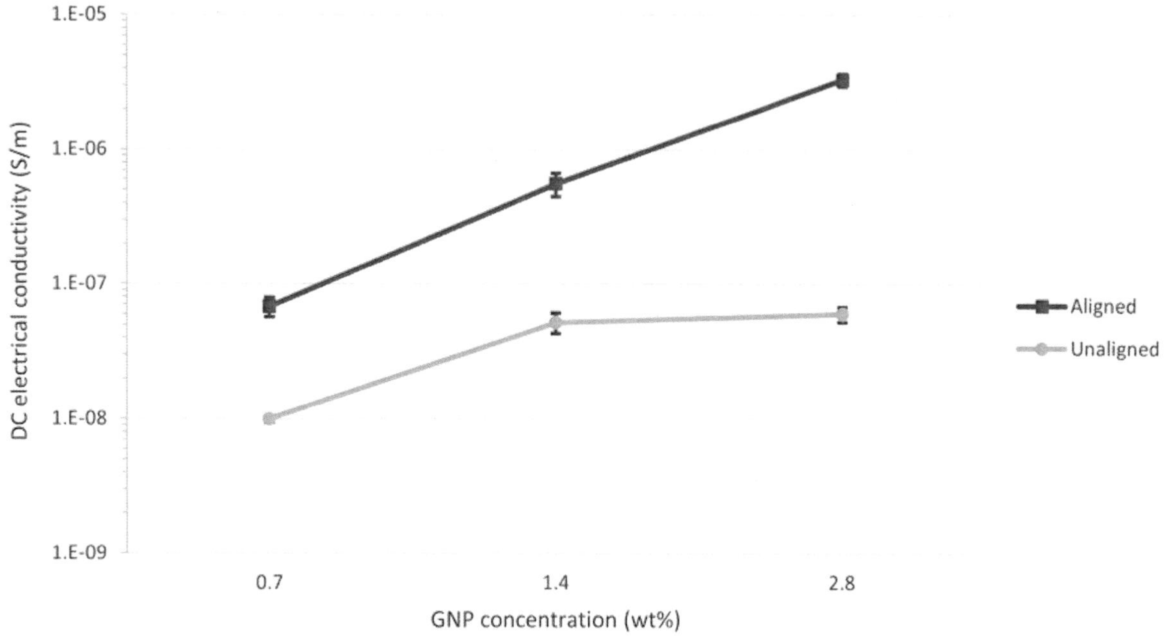

Fig. 11.3 Effects of alignment on DC electrical conductivity for various GNP concentrations

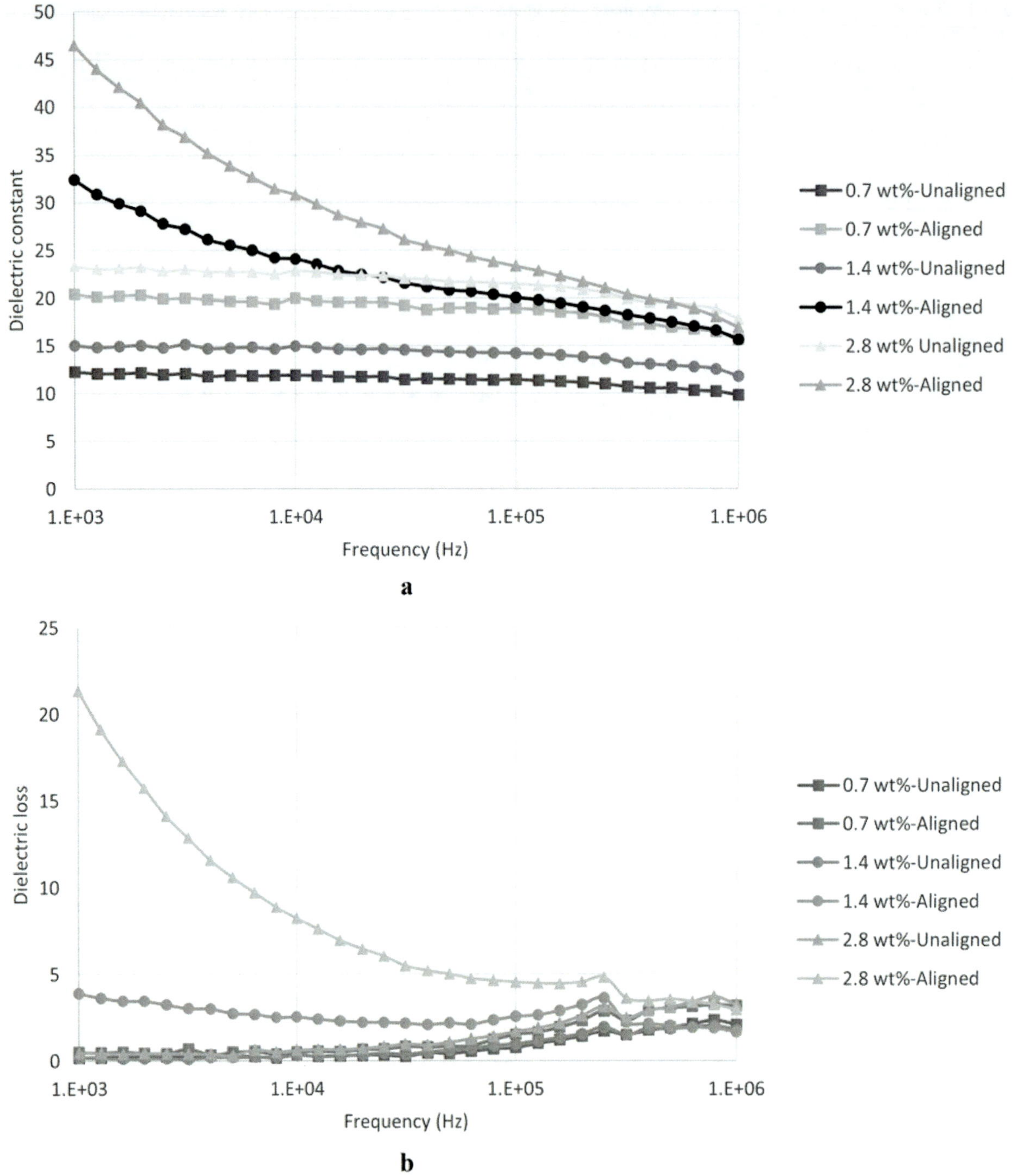

Fig. 11.4 (**a**) Dielectric constant and (**b**) dielectric loss variation of aligned and unaligned epoxy/GNP samples within a frequency range of 10^3–10^6 Hz for different weight concentrations of GNP in epoxy

to the usage of higher weight fractions of GNP, whose translation time is very less compared to 0.1 wt%, which was used by Wu to validate the theoretical model [12]. When the chaining process takes place earlier than rotation, the dipoles get disrupted leading to repulsions of the adjacent particles. To re-establish the chains, more time is required to reorient and successful formation of percolated chains.

The DC electrical conductivity along the aligned direction was calculated for different weight fractions of GNPs in epoxy after curing. Figure 11.3 shows the comparison of DC electrical conductivities at room temperature of aligned and unaligned GNP/epoxy composites. 2–3 order increase in the magnitude of DC conductivity was noticed for aligned samples. This

Table 11.2 Young's modulus increments and DC conductivities of unaligned and aligned epoxy/GNP composites for different weight fractions

GNP concentration (wt %)	% Increase in Young's modulus in aligned direction	Conductivity (S/m) unaligned	Conductivity (S/m) aligned
0.7	7	$1.0 \times 10^{-8} \pm 6.1 \times 10^{-10}$	$6.8 \times 10^{-8} \pm 1.1 \times 10^{-8}$
1.4	9	$5.2 \times 10^{-8} \pm 8.9 \times 10^{-9}$	$5.6 \times 10–7 \pm 1.1 \times 10^{-7}$
2.8	15	$6.0 \times 10^{-8} \pm 7.6 \times 10^{-9}$	$3.3 \times 10–6 \pm 3.6 \times 10^{-7}$

significant increase can be attributed to the chain formation and parallel orientation of the platelets leading to percolated behavior [13]. The electrical percolation threshold of the composites was calculated by using the formula,

$$\sigma = \sigma_0 \left(V_f - V_c\right)^s \tag{11.7}$$

Here, σ is the electrical conductivity of the composite, σ_0 is the electrical conductivity of the GNP, V_f is the volume fraction of the filler, V_c is the percolation threshold, and s is the conductivity exponent. The percolation threshold for randomly oriented samples was 1 wt% whereas for aligned samples it was 0.5 wt%. This demonstrates that the alignment of GNPs reduces the concentration of GNP needed for electrical applications.

Table 11.2 shows the Young's modulus of aligned and randomly oriented samples in the aligned direction of various weight fractions. GNPs have a significantly higher modulus in the parallel direction compared to the perpendicular direction of the platelet. The chain formation tends to accumulate these aligned particles forming a continuous rigid structure of GNPs through the thickness of the epoxy. The increase in Young's modulus compared to randomly oriented samples indicate the mechanical benefits obtained from alignment.

Figure 11.4a shows the dielectric constants calculated over a frequency of 10^3–10^6 Hz for aligned and unaligned samples along the thickness. For unaligned, it is evident that the dielectric constant increases with increase in concentration due to the availability of polarized dipoles in the system [14]. Also, the oblate shape of the particles increases the interfacial polarization (Maxwell-Wagner-Sillars effect) leading to higher charge accumulation [15]. Similar behavior is observed for aligned samples. However, the dielectric constant of the aligned samples shows a significant improvement compared to unaligned samples. This is primarily due to the permanent orientation of the dipoles in the system, percolation, and proximity of the particles, which allows more charge to be accumulated at the interfaces of GNP [16, 17]. However, the unique nature of dielectric curves of 1.4 and 2.8 wt% GNPs in epoxy can be associated with the chain formation which makes the sample relatively more conductive than the 0.7 wt%. This can also be observed in Fig. 11.4b, which shows the dielectric loss of aligned and unaligned GNP/epoxy samples. The dielectric loss significantly increases due to the formation of conductive chains which dissipate the induced stored charge from the electric field. Also, the dielectric constant decreases as frequency increases due to electron hopping taking place at higher frequencies and time for charge accumulation decreases [18].

11.4 Conclusion

Various concentrations of GNPs (0.7, 1.4, and 2.1 wt%) in epoxy were aligned and continuous AC current measurements were used for characterizing the degree of alignment of GNPs. The experimental alignment times for different weight concentrations were compared to theoretical values to assess the importance of viscosity increase due to GNPs. AC conductivity measurements proved to be a practical approach for characterizing the alignment process of GNP/epoxy composite. DC electrical conductivity measurements post curing saw a two- to threefold increase for aligned samples and Young's modulus increment was found to be around (10–15%) compared to unaligned samples. An increase in dielectric constant for a range of frequencies was also noticed for aligned samples due to chaining process, reduction in particle distance due to dipole attraction, and orientation of GNPs. The loss modulus is also increased significantly which indicates the formation of percolated chains.

Acknowledgments This material is based upon work supported by the Basic Research Office, Office of the Under Secretary of Defense for Research and Engineering under award number FA9550-20-1-0156.

References

1. Barber, P., et al.: Polymer composite and nanocomposite dielectric materials for pulse power energy storage. Materials. **2**(4), 1697–1733 (2009)
2. Prolongo, S.G., et al.: In situ processing of epoxy composites reinforced with graphene nanoplatelets. Compos. Sci. Technol. **86**, 185–191 (2013)
3. Potts, J.R., et al.: Graphene-based polymer nanocomposites. Polymer. **52**(1), 5–25 (2011)
4. Keyte, J., Pancholi, K., Njuguna, J.: Recent developments in graphene oxide/epoxy carbon fiber-reinforced composites. Front. Mater. **6**, 224 (2019)
5. Novoselov, K.S., et al.: A roadmap for graphene. Nature. **490**(7419), 192–200 (2012)
6. Papageorgiou, D.G., Kinloch, I.A., Young, R.J.: Mechanical properties of graphene and graphene-based nanocomposites. Prog. Mater. Sci. **90**, 75–127 (2017)
7. Kotikalapudi, S.T., Akula, R., Singh, R.P.: Degradation mechanisms in carbon fiber–epoxy laminates subjected to constant low-density direct current. Compos. Part B. **233**, 109516 (2021)
8. Stankovich, S., et al.: Graphene-based composite materials. Nature. **442**(7100), 282–286 (2006)
9. Wang, F., Wang, H., Mao, J.: Aligned-graphene composites: a review. J. Mater. Sci. **54**(1), 36–61 (2019)
10. Kim, G.H., Shkel, Y.M.: Polymeric composites tailored by electric field. J. Mater. Res. **19**(4), 1164–1174 (2004)
11. Prasse, T., Cavaille, J.-Y., Bauhofer, W.: Electric anisotropy of carbon nanofibre/epoxy resin composites due to electric field induced alignment. Compos. Sci. Technol. **63**(13), 1835–1841 (2003)
12. Wu, S., et al.: Aligning multilayer graphene flakes with an external electric field to improve multifunctional properties of epoxy nanocomposites. Carbon. **94**, 607–618 (2015)
13. Oh Min, K., et al.: Growths of mechanical elasticity and electrical conductance of graphene nanoplatelet/poly (lactic acid) composites under strong electric field: correlation with time evolution of higher order structure of graphene nanoplatelets. Rheol. Acta. **56**(11), 871–885 (2017)
14. Li, Y.C., Tjong, S.C., Li, R.K.Y.: Electrical conductivity and dielectric response of poly (vinylidene fluoride)–graphite nanoplatelet composites. Synth. Met. **160**(17–18), 1912–1919 (2010)
15. Lewis, T.J.: Nano-composite dielectrics: the dielectric nature of the nano-particle environment. IEEJ Trans. Fundam. Mater. **126**(11), 1020–1030 (2006)
16. Xia, X., et al.: Theory of electrical conductivity and dielectric permittivity of highly aligned graphene-based nanocomposites. J. Phys. Condens. Matter. **29**(20), 205702 (2017)
17. Romasanta, L.J., et al.: Functionalized graphene sheets as effective high dielectric constant fillers. Nanoscale Res. Lett. **6**(1), 1–6 (2011)
18. Min, C., et al.: A graphite nanoplatelet/epoxy composite with high dielectric constant and high thermal conductivity. Carbon. **55**, 116–125 (2013)

Chapter 12
How Graphene Oxide Content in Poly (Ethylene Glycol) Affects Phase Separation in Epoxy Matrix for High-Performance Structural Supercapacitor Applications

Alan J. Austin, Sai Tharun Kotikalapudi, Padmanapan Rao, and Raman P. Singh

Abstract Structural supercapacitors represent a promising technology that will help in mitigating range anxiety while providing structural integrity to electric vehicles. In most structural supercapacitors, poly (ethylene glycol) (PEG), the ionic component of structural supercapacitor, is responsible for providing percolating channels for ion transfer. The formation of a percolating network of PEG inside epoxy resin is the result of cure-reaction-induced phase separation (CRIPS) that occurs at curing temperature due to spinodal decomposition. Various nanofillers have been explored (organoclay, TiO_2 nanoparticles, and LLZO) that enhance the ionic conductivity while improving the elastic modulus. Out of these, graphene oxide is a promising nanofiller that has high ionic conductivity due to the presence of functional groups on its basal plane and also provides excellent mechanical properties. Hence, it is instructive to understand the effect of graphene oxide on the size of percolating channels, since these will directly affect the overall performance of the structural supercapacitor. This study explores the effect of change in the weight percentage of graphene oxide on phase separation. To this end, graphene oxide content is varied from 0 to 0.7 weight percentage with respect to PEG to study the magnitude of phase separation between PEG and epoxy. Scanning electron microscopy (SEM) images show promising results suggesting a decrease in pore diameter as amount of graphene oxide is increased, which will enhance the mechanical and electrical properties. By varying the amounts of graphene oxide, we are able to achieve an electrolyte that provides optimal multi-functional performance. A detailed study on the effect of graphene oxide in percolating PEG network in tailoring of mechanical and ionic conductivity properties will be carried out in the near future. These results will be presented and discussed.

Keywords Structural supercapacitors · Phase separation · Graphene oxide · Percolating channel

12.1 Introduction

Efforts have been made to construct a polymer electrolyte with multifunctional applications for a more environmentally friendly and efficient way for storing energy in structural supercapacitors. The electrolyte must provide sufficient mechanical strength for load-bearing applications as well as provide optimal ionic conductivity. Poly (ethylene oxide) (PEO) was found to be able to coordinate lithium ions due to its ether oxygen (EO) repeating unit [1, 2]. Since this discovery, much research has been dedicated to finding a suitable PEO-based electrolyte. However, PEO has insufficient ionic conductivity for a structural supercapacitor at ambient temperature [3]. Gel polymer electrolytes have high ionic conductivity, but lack the

A. J. Austin (✉) · P. Rao
School of Materials Science and Engineering, Oklahoma State University-Tulsa, Tulsa, OK, USA
e-mail: alan.austin@okstate.edu; padmanapan.rao@okstate.edu

S. T. Kotikalapudi
School of Mechanical and Aerospace Engineering, Oklahoma State University, Stillwater, OK, USA
e-mail: skotika@okstate.edu

R. P. Singh
School of Materials Science and Engineering, Oklahoma State University-Tulsa, Tulsa, OK, USA

School of Mechanical and Aerospace Engineering, Oklahoma State University, Stillwater, OK, USA
e-mail: raman.singh@okstate.edu

V. Chalivendra, F. Gardea (eds.), *Mechanics of Composite, Hybrid & Multi-functional Materials, Volume 5*, Conference Proceedings of the Society for Experimental Mechanics Series, https://doi.org/10.1007/978-3-031-17445-2_12

mechanical properties for multifunctional applications. To achieve multifunctional performance from a polymer electrolyte, a one polymer system lacks either the mechanical or ionic properties for a structural supercapacitor. A binary polymer system consisting of a liquid ionic component and a structural polymer will allow for multifunctional performance.

One such binary system consists of PEG and epoxy. Poly (ethylene glycol) (PEG) at lower molecular weight has the same repeating ether oxygen monomer unit as PEO, which allows for the same high affinity for lithium-ion coordination [4]. PEG 600 molecular weight is a soft solid at ambient temperature which allows for high ionic mobility, and allows for preparing the electrolyte without having to use a volatile solvent. Epoxy resin is a commonly used thermosetting polymer that has excellent mechanical and thermal properties, and is very common in many structural applications. While initially miscible, PEG and epoxy will phase separate by spinodal decomposition when curing reaction occurs, and create a co-continuous morphology which will allow for ionic channels to percolate through the non-conductive epoxy matrix [5]. This will allow for multifunctionality of the matrix when the SPE is adhered to the carbon fiber of the structural supercapacitor [6]. However, when ionic channels are created through the epoxy matrix it is usually at the cost of degrading its mechanical properties [7].

Incorporating nanofillers is an effective way to control the final morphology and phase separation of an epoxy-based polymer electrolyte [8]. Graphene oxide (GO) is a promising nanofiller that has excellent mechanical and thermal properties, and contains multiple oxygen-rich functional groups (epoxy, hydroxyl, and carboxyl) [9]. These functional groups break up the strong affinity of the repeating EO monomer unit of PEO and lithium-ion, and free up more lithium-ions for ion transport allowing for better salt dissociation [10]. Research has shown that GO can create continuous ion channels within polymer electrolytes [11]. This study will show how graphene oxide content (0–0.7 weight percentage with respect to PEG) effects phase separation between PEG and epoxy, and how it effects the mechanical properties for future structural supercapacitor applications.

12.2 Experimental

12.2.1 Materials and Methods

The ionic polymer incorporated was PEG with an average molecular weight (600) which was purchased from Acros Organics™. Graphene oxide was obtained from Nanoshel® UK LTD. The epoxy resin which was used in this study is EPON™ Resin 862 and EPIKURE™ Curing Agent 3724 was from Hexion Inc.

PEG was weighed and added into a clean beaker, and heated to 30 °C until it becomes a liquid. GO was then weighed from 0.1% to 0.7% with respect to PEG (0.021–0.143 g), and added into the liquid PEG. To ensure we achieve a uniform dispersion of GO in PEG the following steps are taken [8]: The PEG and GO mixture were stirred vigorously using magnetic stir for approximately 1 h. Shear mixing (ULTRA-TURREX® T 25 basic) is then followed at 11,000 rpm for 5 min. While shear mixing, we made sure to keep it in an ice bath to avoid overheating that could possibly damage the functional groups of GO. The last step for proper dispersion of GO was bath sonication (VWR® Model 75D) for 30 min. Epoxy is then poured into a separate beaker, and is mixed with the hardening agent in a 5:2 ratio. It was then degassed to remove the bubbles caused by hand mixing. The PEG/GO ionic component is then poured into the epoxy/hardener structural component, and was mixed by magnetic stir for approximately 45 min to ensure a homogeneous mixture. The electrolyte was then degassed and poured into a silicone mold to cure at 60 °C in a vacuum oven (Across International) for 24 h.

12.2.2 Characterization

After the SPE is completely cured, a 4.8 mm (3/16″) diameter cutting tool was used to cut five circular samples. The initial weight was taken and the samples were placed in a 50:50 water-alcohol solution, and then bath sonication (power level 3) was used for 2 h. Once bath sonication is complete, the cut samples were then submersed in water for a minimum of 16 h. The specimens were then placed in a vacuum oven until the mass of the samples was constant and no more weight was being lost. The specimens were then sputtered (Leica EM ACE600) with Iridium Han for 80 s. The scanning electron microscope (Hitachi S-4800) was used. The accelerating voltage was set to 15 kV and the emission current was set at 10 μA.

The SPE was made using the same ratios and procedure as previously mentioned, and was prepared for mechanical characterization. ASTM D695 was used for characterizing the compressive strength of the solid polymer electrolyte. For carrying out the ASTM standard, we used Instron 5567 equipped with a 10 kN load cell. Specimens were poured into a silicone mold in the dimensions of 12.7 × 12.7 × 25.4 mm (rectangular prism). Five identical specimens were made for each compression test as called for by the standard and the average was taken. Each specimen was then aligned directly in the center of the two platelets and a compressive force (F_c) was then applied. The standard speed was set at 1.3 mm/min. The compressive stress (σ_c) and compressive strain (ϵ_c) were calculated using the following equations:

$$\epsilon_c = \frac{\Delta l}{l} \tag{12.1}$$

$$\sigma_c = \frac{F_c}{A} \tag{12.2}$$

Here, the compressive strain is calculated by dividing the change in length (Δl) by the original length (l), and the compressive stress is found by dividing the compressive force divided by the area of the specimen. The compressive modulus (E_c) was then calculated by taking the slope of the linear portion of the stress-strain curve between 2% and 10% strain.

12.3 Results

Table 12.1 shows the weight of SPE before and after the PEG was removed. The expected weight loss should be approximately 50% as the samples were made with a 50:50 weight fraction of PEG and epoxy. PEG can be removed from the epoxy matrix using a polar solvent (water and alcohol mixture) since PEG is hydrophilic. This would indicate that PEG and epoxy were in two separate phases, and that ionic channels were created within the epoxy matrix. The weight loss decreased with increasing amounts of GO, but at 0.4 and 0.5 weight percent we saw a slight increase, which could be caused by removing GO since it is hydrophilic, or the pores became more percolated in the epoxy matrix. However, at 0.6 and 0.7 weight percent of GO the weight loss of PEG decreases. Using higher amounts of GO created excess aggregations embedded in the polymer matrix indicating that GO was not well dispersed, or with increasing amounts of GO that the ionic channels were not fully percolated through the epoxy matrix.

Figure 12.1 shows the micrographs of GO incorporated into the PEG-epoxy electrolyte, and its effects on the final microstructure. The SEM images when there is no GO present in the 50:50 PEG-epoxy binary mixture depict a porous microstructure where the PEG has been removed from the epoxy matrix. This indicates that PEG was interpenetrated through the cross-linked epoxy network, where the bright region is the epoxy phase that is left and leaving a bicontinuous microstructure where the epoxy nodules show a connected-sphere morphology, and as a result is favorable for ionic conductivity due to the channels generated [9]. When 0.3% GO is incorporated into the electrolyte the resulting morphology shows a homogeneous single-phase, and not a porous structure like the images above [12]. This homogeneous single-phase has pores that become increasingly smaller and are randomly dispersed in the epoxy matrix. GO content at 0.3% looks is well dispersed in the polymer matrix, and minimal aggregations would potentially interfere with ion transport [10]. However, we lose the disordered porous microstructure that was generated in the SEM images without GO. This is indicative that phase separation was possibly suppressed due to the viscosity increase caused by GO [14]. The SEM images of when 0.6% GO content is added show large aggregations caused from GO not being well dispersed in the polymer matrix [8]. These aggregations could possibly hinder ion transport, thus hurting the potential of ionic conductivity [13].

The results of the mechanical characterization with increasing GO amount (0–0.7 weight % with respect to PEG) in 50:50 weight fraction PEG-epoxy mixture are shown in Table 12.2. When GO is not incorporated in the electrolyte, the resulting morphology is typical of a liquid-rich phase-separated system [3]. When GO (0.1–0.5% GO) is added, the morphology changes from a bicontinuous network to a single-phase material where small size pores are randomly distributed throughout

Table 12.1 Weight loss% of 50:50 PEG epoxy mixture with varying GO content (0–0.7%)

GO weight (%)	Initial weight (mg)	Final weight (mg)	Weight loss (%)
0	44.7	22.6	49.4 ± 1.7
0.1	44.6	23.5	47.7 ± 1.5
0.2	47.6	25.1	47.3 ± 1.9
0.3	46.3	24.3	47.5 ± 2.1
0.4	48.5	25.1	48.2 ± 1.1
0.5	49.4	25.4	48.6 ± 1.3
0.6	48.8	27.1	44.5 ± 2.6
0.7	52.5	29.3	44.2 ± 2.8

(a)

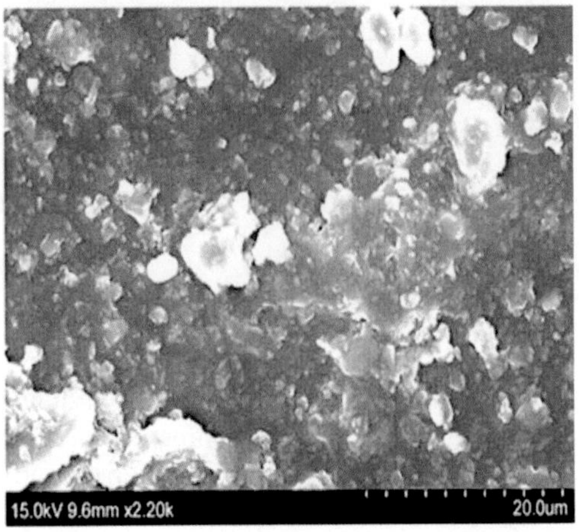

(b)

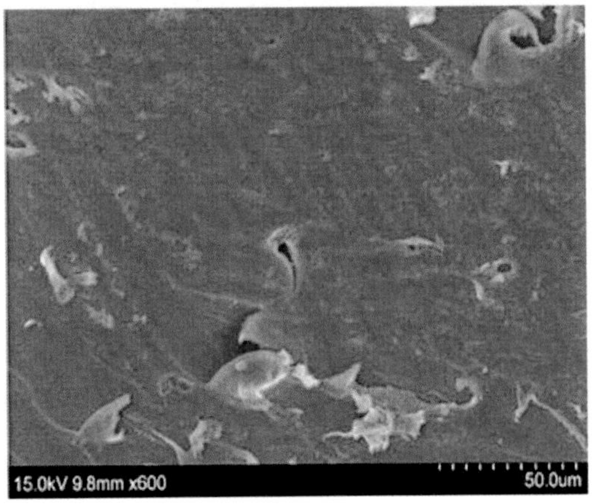

(c)

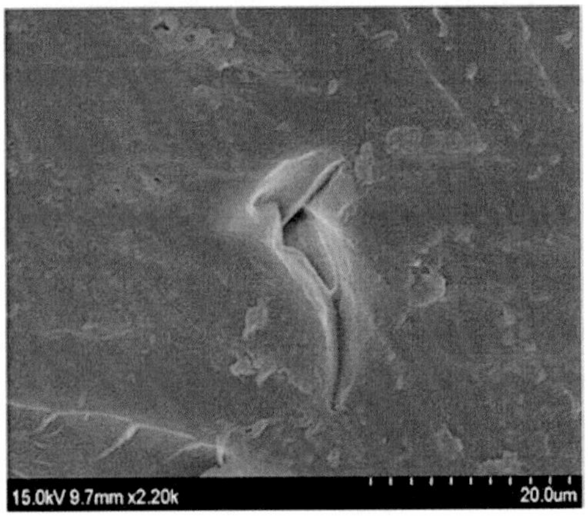

(d)

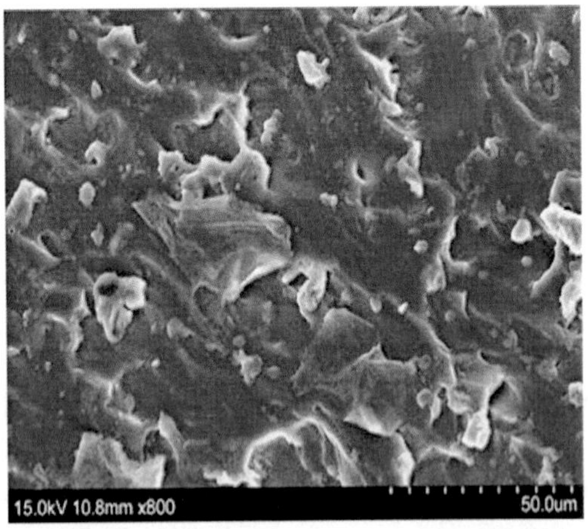

(e)

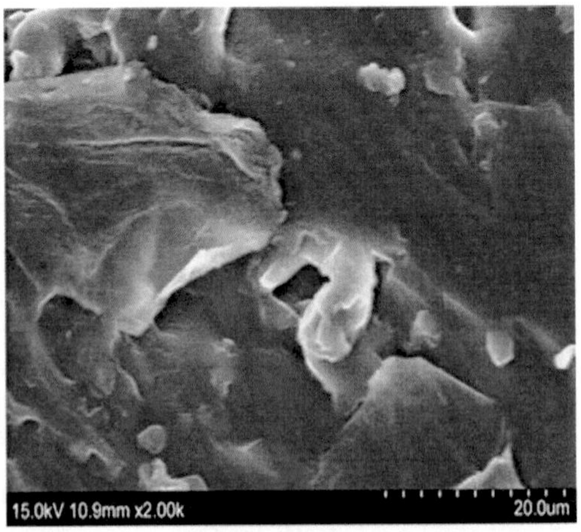

(f)

Fig. 12.1 SEM images of varying GO content at different magnifications (**a**) 0% GO 50 μm, (**b**) 0% 20 μm, (**c**) 0.3% GO at 50 μm, (**d**) 0.3% GO at 20 μm, (**e**) 0.6% GO at 50 μm, and (**f**) 0.6% at 20 μm

Table 12.2 Compression testing values of 50:50 PEG-epoxy mixture with varying GO content (0–0.7%)

GO weight (%)	E_c (MPa)	E_c (%) Increase	Max. σ_c (MPa)	σ_c (%) Increase
0	0.78 ± 0.03	–	0.42 ± 0.06	–
0.1	0.91 ± 0.03	16.7	0.62 ± 0.06	47.6
0.2	0.95 ± 0.02	21.8	0.63 ± 0.08	50.0
0.3	0.97 ± 0.01	24.4	0.54 ± 0.03	28.6
0.4	1.01 ± 0.04	29.5	0.55 ± 0.05	30.9
0.5	1.04 ± 0.04	33.3	0.54 ± 0.05	28.6
0.6	1.11 ± 0.05	42.3	0.53 ± 0.07	26.2
0.7	0.89 ± 0.03	14.1	0.54 ± 0.04	28.6

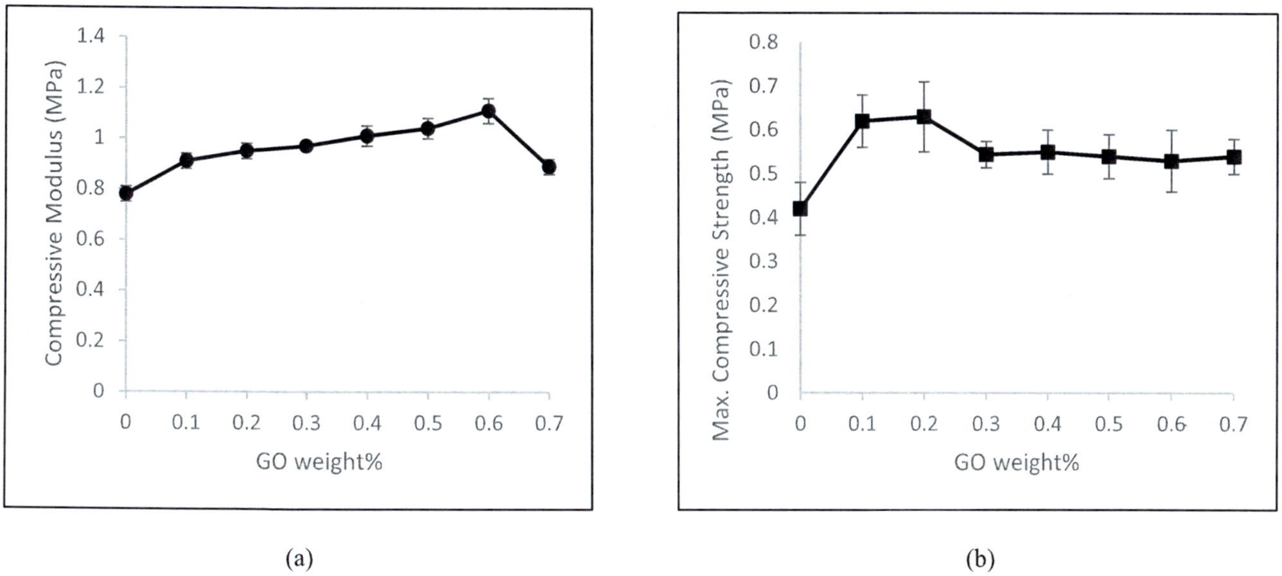

Fig. 12.2 (**a**) Compressive modulus of 50:50 PEG epoxy mixture with varying GO content (0–0.7%). (**b**) Maximum compressive strength with varying GO content (0–0.7%)

the non-conductive epoxy matrix [9]. This morphology resulted in improvement of both the compressive modulus and strength. The compressive modulus shows a linear progression of increasing, and the highest compressive modulus was recorded at 0.6% GO. This is a 42.3% increase in compressive modulus when there is 0% GO present. When GO is at 0.7% with respect to PEG, the compressive modulus declines due to agglomeration of GO embedded in the polymer matrix. Further research is being conducted to understand why the compressive modulus declines at 0.7% GO. The compressive strength peaked at 0.2% GO and is a 50% increase in maximum compressive strength, and then the compressive strength drops and stays consistent from GO content 0.3–0.7% (Fig. 12.2).

12.4 Conclusion

This study takes a look into incorporating GO into a binary polymer electrolyte with a 50:50 weight percent of structural and conductive phase, and its effects on phase separation and structural properties. The final microstructure has a significant impact on the performance of a polymer electrolyte, and GO allows for tailoring of the microstructure to achieve multifunctional performance. Introduction of GO into the PEG/epoxy binary mixture changed the microstructure from a bicontinuous two-phase morphology to a more plasticized single phase where the pores become increasingly smaller. Thus, the addition of GO made PEG more miscible in epoxy when cross-linking reactions occur. The electrolyte shows improvement in structural properties when GO is incorporated. An increase in maximum compressive strength up to 50% was achieved at 0.2% GO content compared to when no GO was added in the 50:50 PEG epoxy mixture, and the compressive modulus was increased up to 42.3% at 0.6% GO compared to no GO in the 50:50 PEG epoxy mixture. GO content from 0.4% to 0.5% with respect to PEG allows for good dispersion in the polymer matrix, and shows improvement in structural properties. PEG was extracted from the epoxy matrix at a higher percentage than the rest of the samples with GO added, and could indicate that the ionic channels created were more percolated. GO between 0.4% and 0.5% is promising for an epoxy-based polymer electrolyte for future structural supercapacitor applications.

Acknowledgments This material is based upon work supported by the Basic Research Office, Office of the Under Secretary of Defense for Research and Engineering under award number FA9550-20-1-0156.

References

1. Barrett, C.R., Nix, W.D., Nix, W.D., Tetelman, A.S., et al.: The Principles of Engineering Materials. Prentice Hall (1973)
2. Fenton, D.E., Parker, J.M., Wright, P.V.: The conductivity in peo-based polymer electrolytes. Polymer. **14**, 589–591 (1973)
3. Gienger, E.B., Nguyen, P.-A.T., Chin, W., Behler, K.D., Snyder, J.F., Wetzel, E.D.: Microstructure and multifunctional properties of liquid+ polymer bicomponent structural electrolytes: epoxy gels and porous monoliths. J. Appl. Polym. Sci. **132**(42), 42681 (2015)
4. Andrews, W.T., Liebig, A., Cook, J., Marsh, P., Ciocanel, C., Lindberg, G.E., Browder, C.C.: Development of a peo-based lithium ion conductive epoxy resin polymer electrolyte. Solid State Ionics. **326**, 150–158 (2018)
5. Zhang, R., Zhang, L.: Preparation of 3-dimentional skeletal polymer via control of reaction-induced phase separation in epoxy/poly (ethylene glycol) blends. Polym. Bull. **61**(6), 671–677 (2008)
6. Fu, Y., Zhou, H., Zhou, L.: Phase-microstructure-mechanical properties relationship of carbon fiber reinforced ionic liquid epoxy composites. Compos. Sci. Technol. **207**, 108711 (2021)
7. Schneider, L.M., et al.: Bicontinuous electrolytes via thermally initiated polymerization for structural lithium ion batteries. ACS Appl. Energy Mater. **2**(6), 4362–4369 (2019)
8. Yalin, Y., Zhang, B., Wang, Y., Qi, G., Tian, F., Yang, J., Wang, S.: Co-continuous structural electrolytes based on ionic liquid, epoxy resin and organoclay: effects of organoclay content. Mater. Des. **104**, 126–133 (2016)
9. Zhongliang, H., Zhang, X., Liu, J., Zhu, Y.: Ion liquid modified go filler to improve the performance of polymer electrolytes for li metal batteries. Front. Chem. **8**, 232 (2020)
10. Yuan, M., Erdman, J., Tang, C., Ardebili, H.: High performance solid polymer electrolyte with graphene oxide nanosheets. RSC Adv. **4**(103), 59637–59642 (2014)
11. Mohammadsalih, Z.G., Inkson, B.J., Chen, B.: The effect of dispersion condition on the structure and properties of polystyrene/graphene oxide nanocomposites. Polym. Compos. **42**(1), 320–328 (2021)
12. Kwon, S.J., Kim, T., Jung, B.M., Lee, S.B., Choi, U.H.: Multifunctional epoxy-based solid polymer electrolytes for solid-state supercapacitors. ACS Appl. Mater. Interfaces. **10**(41), 35108–35117 (2018)
13. Shim, J., Kim, D.-G., Kim, H.J., Lee, J.H., Baik, J.-H., Lee, J.-C.: Novel composite polymer electrolytes containing poly (ethylene glycol)-grafted graphene oxide for all-solidstate lithium-ion battery applications. J. Mater. Chem. A. **2**(34), 13873–13883 (2014)
14. Guijun, Y., Peiyi, W.: Effect of chemically modified graphene oxide on the phase separation behaviour and properties of an epoxy/polyetherimide binary system. Polym. Chem. **5**(1), 96–104 (2014)